AF361310

# Praise for *New Approaches to Contemporary Adaptation*

"*New Approaches to Contemporary Adaptation* is very admirable, an apt representation of where adaptation studies can travel once scholars, readers, viewers, and artists have released adaptations and sources from binary ways of thinking. The breadth of the volume and its insistence on breaking out of traditional concepts will intrigue readers across disciplines."

> —Julie Grossman, professor of literature and film studies
> at Le Moyne College and co-editor of the Adaptation and
> Visual Culture series with Palgrave Macmillan

"This is a new phase adaptations book, one that adopts an open and inclusive approach to adaptation studies. Complementing recent works in the field, Kaklamanidou's book broadly understands the process of adaptation as the audio-visualization of words and images. The result is a vibrant collection of chapters, which ranges across literature and music, film and television to demonstrate the urgency of adaptation studies in the humanities today."

> —Constantine Verevis, Monash University

"True to its title, *New Approaches to Contemporary Adaptation* offers new insights into a well-established field. Leaning into her structuralist training, Betty Kaklamanidou, as both a writer and an editor, works to redefine 'adaptation,' and, in the process, to extend the scope of adaptation studies. This collection concerns itself with adaptations in broad cultural, linguistic, musical, and genre contexts, including well-researched chapters on Portuguese and Greek adaptations. The contributors to the collection do their homework, and the chapters on long-form television adaptations are particularly impressive."

> —Dennis Cutchins, co-editor of *The Routledge Companion to Adaptation*

"*New Approaches to Contemporary Adaptations* calls for the development of new ways of looking at adaptations and paves the way for future work in the field by highlighting the importance of data compilation and analysis. The anthology brings together ten fascinating contributions which illuminate its theoretical framework and reflect the increasingly diverse landscape of adaptations and their study."

> —Costas Constandinides, author of *From Film
> Adaptation to Post-celluloid Adaptation*

# NEW APPROACHES TO CONTEMPORARY ADAPTATION

# Contemporary Approaches to Film and Media Series

*General Editor*

Barry Keith Grant, Brock University


*A complete listing of the books in this series can be found online at* wsupress.wayne.edu

# NEW APPROACHES TO CONTEMPORARY ADAPTATION

*Edited by*
BETTY KAKLAMANIDOU

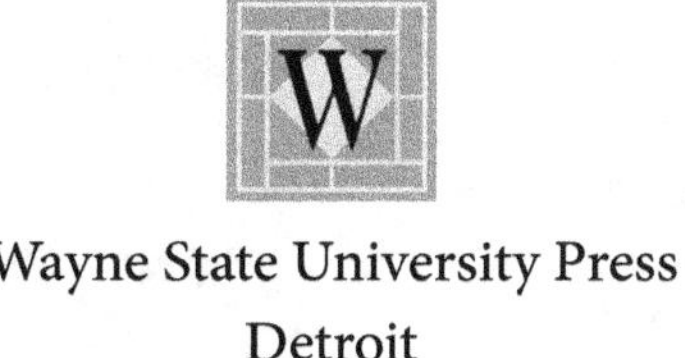

Wayne State University Press

Detroit

ISBN 978-0-8143-4625-9 (paperback); ISBN 978-0-8143-4624-2 (case);
ISBN 978-0-8143-4626-6 (ebook)

Library of Congress Control Number: 2020938189

Wayne State University Press
Leonard N. Simons Building
4809 Woodward Avenue
Detroit, Michigan 48201-1309

Visit us online at wsupress.wayne.edu

To Laurence Raw

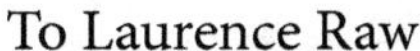

You'll always be part of this collection

# Contents

# Preface

Betty Kaklamanidou

THE PRELIMINARY REMARKS OF this edited collection begin with the following brief personal account—let's call it a micro-historical "adaptation" of my almost twenty-year liaison with the field (actually one of my longest relationships). It was in the summer of 1999 that my relationship with adaptation started in a rather unorthodox way. I remember precisely the moment I told my friend at a coffee shop, having just watched *Cruel Intentions*, that I had decided to use *Les Liaisons Dangereuses* and its four filmic adaptations—thus far—as my PhD case studies. I was convinced that had Choderlos de Laclos lived in the twentieth century, his novel would surely resemble Roger Kumble's version more than Stephen Frears's in 1988 and Miloš Forman's in 1989 and would certainly have nothing to do with Roger Vadim's version in 1959. Little did I know that my utter conviction about Kumble uncovering the "essence" of Laclos's masterpiece coincided with most writing regarding the infamous fidelity issue that still accompanies adapted film texts.[1] Of course, I had to convince my thesis supervisor that the epistolary novel's cinematic adaptations were chosen not out of personal and therefore subjective preference but based on specific criteria. I therefore created a corpus of around three thousand film adaptations produced in the United States and the United Kingdom from 1914 to 1996, analyzed them statistically, and proved that Laclos is one of the rare eighteenth-century writers—in fact, he was in an elite group of just five eighteenth-century writers, among the total of 632 writers, 554 of whom were twentieth-century novelists—whose work was adapted in the twentieth century.

I began my dissertation in 2000, a time when the major scholarly works on adaptation were George Bluestone's *Novel and Film*,[2] Brian McFarlane's *Novel to Film*,[3] Deborah Cartmell and Imelda Whelehan's anthology *Adaptations: From Text to Screen, Screen to Text*,[4] and a few seminal journal articles—mind you, research in the late 1990s in Greece was radically different from today; certainly the Internet was not as helpful back then. The scarcity of sources for a PhD candidate, however, meant I had the opportunity to offer a small contribution to the field with greater ease. Indeed, my structuralist/semiotic "upbringing" in academia assisted me in combining Algirdas J. Greimas's narrative grammar and Gérard Genette's transtextual categorization and theory of focalization and apply them to my case studies. From 2006 onward, I began teaching a course on literature and film, while adaptation studies was beginning to amass an abundance of publications.

In 2017, I organized an adaptation conference in Greece. On the one hand, the plethora of new contextual approaches, as well as the new angles from which colleagues treated adapted texts, was exhilarating. On the other hand, some of the questions I had been wondering and reading about over the course of eighteen years had yet to be answered. These unanswered questions led to this collection, whose introduction is an effort to offer further areas of exploration to future adaptation scholars.

## Notes

1. Among others, see William Luhr, "Adapting *Farewell, My Lovely*," in *A Companion to Literature and Film*, ed. Robert Stam and Alessandra Raengo (Malden, MA: Blackwell, 2004), 278–97; Jack Boozer, "Introduction: The Screenplay and Authorship in Adaptation," in *Authorship in Film Adaptation*, ed. Jack Boozer (Austin: University of Texas Press, 2008), 1–30; and Thomas Leitch, *Film Adaptation and Its Discontents* (Baltimore: Johns Hopkins University Press, 2011).

2. George Bluestone, *Novels into Film* (Baltimore: Johns Hopkins University Press, 1957).

3. Brian McFarlane, *Novel to Film* (Oxford: Clarendon Press, 1996).

4. Deborah Cartmell and Imelda Whelehan, eds., *Adaptations: From Text to Screen, Screen to Text* (New York: Routledge, 1999).

# Acknowledgments

It was Friday the 13th back in April 2018 and I was enjoying my coffee in Thessaloniki with a Greek colleague from the University of Alabama who was in the city for her Easter break. We were having a great time, talking about our academic adventures and future plans until the sound of an incoming email on my smartphone interrupted us. Glancing over at the sender's name and the subject, I asked my friend to read it for me and tell if it was good or bad news. It was the former. I had just been offered an advance contract by Wayne State University Press for the adaptation collection I had envisioned the year before. I hugged my friend (back at a time when we could still hug each other), thanking her for bringing me such good luck and treated us both to some wonderful sweets.

Fast forward to June 2020. After three blind peer reviews, rewrites, updates, revisions and corrections, proofreading, and editorial decisions the collection has reached its final production stage.

It is at this time I'd like to thank all the people who played a part in the final shape of this volume. First in the list is Marie Sweetman, my Wayne State University Press editor who believed in the project from the beginning, always replied within hours, had great suggestions, and a wonderful disposition at all times. I'd also like to thank my three reviewers for their suggestions for revisions, their thoughts, and the time they took to evaluate the proposal. They underlined a couple of important references I might have missed and helped me improve the collection's structure and tighten my main argument. Special thanks to WSUP series editor Barry Keith Grant for his support of the collection. His recommendation and his suggestions resulted in a stronger version of the anthology. Finally, I'd like to acknowledge all the great people at WSUP, who were always there to assist and provide useful pieces of advice.

My heartfelt appreciation goes to my ten contributors. Having to put up with my editing methods can be annoying but the authors were nothing but polite and punctual and, most importantly, delivered stimulating chapters that taught me a lot while expanding the field of adaptation studies. Thomas, Eurydice, Christina, Simon and Stacey, Karin, Ursula-Helen, Nicole, Joakim, and Thomas: Thank you!

I wouldn't be where I am today without three people, who despite not knowing anything about "adaptation studies" have actually shaped part of this collection because they have shaped parts of me. This is for you Petros, Mom, and Lia!

# Introduction

## Inclusivity and Possibilities

### Betty Kaklamanidou

*I'd like to thank Margaret Tally and Thomas Leitch for taking the time to read this introduction and offer valuable insights and suggestions.*

WHAT CAN YET ANOTHER collection on adaptation offer in 2020? Is it even possible to offer something entirely new? The persistent problem of the "already said"—posed by Umberto Eco[1]—as well as the wealth of academic work on the subject certainly complicate matters for contemporary scholars. In other words, if you are looking for something that hasn't existed before in adaptation studies, set your book or screen aside and go for a walk. Yet if you are looking for new thoughts on old issues as well as new insights on recent developments, then continue reading. The aim of this anthology is to open up the field of adaptation studies by revisiting the object of analysis and proposing alternative ways of looking at it.

In *The Oxford Handbook of Adaptation Studies*, editor Thomas Leitch provides an excellent and concise account of the history of adaptation studies, dividing it into four periods.[2] The first is the "prehistoric period," or "Adaptation Studies 0.0," which culminates with the publication of George Bluestone's seminal *Novels into Film* in 1957; the second, "Adaptation Studies 1.0," offers the basic principles of adaptation (i.e., Geoffrey Wagner's and Dudley Andrew's categorizations) in a series of books that include paradigmatic case studies and ends with Brian McFarlane's 1996 *Novel to Film*; the third, "Adaptation Studies 2.0," delves into the examination of pop culture and offers more sophisticated analytical and evaluative methods. This period is introduced by Deborah Cartmell and Imelda Whelehan's 1999 anthology, followed by Robert Stam's edited collection with Alessandra Raengo and his subsequent monograph[3] as well as other significant publications; the fourth period, "Adaptation Studies 3.0," is characterized by its embrace "of digital technologies" and its "suspicion of the limits of intertextuality as a methodological framework."[4] Although Leitch does not mention a specific book of this period, we could safely hypothesize that Costas Constandinides's *From Film Adaptation to Post-Celluloid Adaptation*,[5] Kathleen Loock and Constantine Verevis's *Film Remakes, Adaptations and Fan Productions*,[6] and Carolyn Jess-Cooke and Constantine Verevis's *Second Takes: Critical Approaches to the Film Sequel*[7] best exemplify this period, alongside several monographs and anthologies that focus on either canonical literature or specific writers and films.

Writing during this fourth period and taking into account the more than often long period it takes for an academic collection to reach publication from initial inception (not to mention that Leitch's anthology contains an impressive forty-two chapters), the editor was by definition not able to keep up with the latest relevant publications or foresee what would happen in the immediate future. In 2015, Palgrave inaugurated its Palgrave Studies in Adaptation and Visual Culture series. The first book, Julie Grossman's *Literature, Film, and Their Hideous Progeny: Adaptation and ElasTEXTity*,[8] is—only four years later—accompanied by seventeen titles, whose subject varies from examining television and adaptation (Yvonne Griggs's *Adaptable TV*)[9] and childhood and adaptation (Robyn McCallum's *Screen Adaptations and the Politics of Childhood*)[10] to focusing on a single fictional character and his multi-platform adaptations (Jeremy Strong's edited collection *James Bond Uncovered*)[11] and national-specific adaptations (Petr Bubeníček's study on adaptations in communist Czechoslovakia in *Subversive Adaptations*).[12] Between 2015 and 2017, the field of adaptation studies witnessed an impressive wealth of publications focusing on old and new paths scholars decided to take. From Iain Robert Smith's *The Hollywood Meme: Transnational Adaptations in World Cinema*[13] and Dru Jeffries's *Comic Book Film Style: Cinema at 24 Panels per Second*,[14] which tackled non-Anglophone films and sources outside the novel, to the more theoretical yet equally timely and necessary discussions in Grigg's *The Bloomsbury Introduction to Adaptation Studies: Adapting the Canon in Film, TV, Novels and Popular Culture*,[15] Julie Grossman and R. Barton Palmer's edited collection *Adaptation in Visual Culture: Images,*

*Texts, and Their Multiple Worlds*,[16] and several other recent titles, there is an exponential increase in the literature, followed by a great number of books that study a specific filmmaker, author, or literary genre (among them, Douglas McFarland and Wesley King's *John Huston as Adaptor*,[17] Richard J. Hand and Andrew Purssell's *Adapting Graham Greene*,[18] Wieland Schwanebeck and Douglas McFarland's *Patricia Highsmith on Screen*,[19] and Max Sexton and Malcom Cook's *Adapting Science Fiction to Television*).[20]

This cornucopia of relevant literature[21] that began in 2015 has ushered in the fifth historical period of adaptation studies, namely, "Adaptation Studies 4.0," to follow Leitch's historical division. The previous periods have provided us with a wealth of methodologies, theoretical concepts, and sophisticated paradigms of case studies analyses, while this era exploits the past academic heritage and also expands the field to new and exciting realms. Therefore, this collection aims to act as a complement to this phase of adaptation studies. It tries to propose answers to the following persistent questions while also addressing some new queries:

> What is the definition of an adaptation? What is necessary for an adaptation to be treated as such only if the source acknowledges its status as a derivative work? What happens with films that are based on "true stories"? Aren't they adaptations of the real events? Didn't the filmmakers base their fictional narrative on newspaper articles, archival material, and witnesses' interviews (when possible) in the same way a screenwriter reimagines a novel?

> What kinds of stories have been adapted in the last couple of decades? The absence of concrete data, even originating from the Hollywood paradigm, may lead to erroneous conclusions. Is the novel still the primary source in adaptation? What is the percentage of comics, graphic novels, children's books, and historical documents that lend their stories to film?

> How do transnational adaptations differ from adaptations that target the same national audiences as the texts they adapt? For instance, how does Greek cinema adapt a Hollywood musical? Why do adaptation studies predominantly examine Anglo-American film productions and sources?

> What do television serials that last for several seasons actually adapt
> when their source is a single book or novel (or both)? How do
> ongoing popular television shows tackle the issue of a finite or
> incomplete source?
>
> How do extratextual parameters, such as producers' input, censor-
> ship, and specific sociopolitical situations, affect the adaptation
> process?
>
> What happens when we change the focus to examining the visuality,
> inherent in some relatively new sources (comics, graphic novels,
> video games), and how does it influence film adaptations?

## Proposing an Inclusive Perspective for Adaptations

Each scientific field involves a clearly demarcated object of study. Yet, humanities and more specifically adaptation studies often face obstacles in this delineation process. Adaptation studies grew initially out of an academic interest in literature and film. This relationship evolved with the inclusion of relevant university courses, the establishment of academic journals, and the organization of conferences. Film and literature became the focus of adaptation studies, obscuring at the same time other forms, such as stage adaptations, television adaptations, musical adaptations, and so forth, not to mention a host of e-adaptations in the era of Web 2.0. Thomas Leitch already addressed this "limitation" in 2012, underlining that the emphasis on the transformation of literary texts into cinematic moving images provided "two possible disciplinary homes . . . departments of literature and of cinema."[22]

This collection includes a very interesting chapter on experimental musical adaptations based on a variety of sources (from samples to pure imagining). This chapter was added to the collection precisely to suggest that other fields (i.e., music studies) could profitably offer their theories, methodologies, and contexts regarding the practice of adaptation that might inspire younger colleagues to follow suit in their publication journeys.

The object of adaptation studies seems fairly simple: *The object of adaptation studies is adaptations—that is, films based on stories found in other media.*

This, at first glance, uncomplicated and seemingly self-evident formulation includes one main assumption and invites three questions if one looks more closely. The assumption, already hinted at in the title of this section, is that I take film adaptation as the object of adaptation studies without further explanation. However, this seemingly exclusionary decision is a pragmatic and not a theoretical result. In other words, film adaptations are indeed the object of study for the specific part of adaptations studies that examines how this medium uses, relates to, takes advantage of, and is influenced by each source. In this way, I expect the future to bring more branches of adaptation studies that focus on different adaptations to the foreground. We could, therefore, have ballet adaptation studies, radio adaptation studies, video game adaptation studies, music adaptation studies, opera adaptation studies, and so on. In this way, concurring with and daring to expand upon James Naremore's conclusion regarding how adaptation can "move from the margin to the center of contemporary media studies,"[23] I envision adaptation studies as an umbrella discipline that can host a variety of branches and find it is time to move away from the prevalent film/source paradigm and examine how adaptation works when the final product is not a movie.

The three questions that emerge from the answer above concerning the object of film adaptation studies are as follows:

a. Do "other media" include real-life stories, whether or not they circulate in another, most often written form (memoirs, historical documents, newspaper/online articles)? What about stories that can be found in other places, such as memories? Greta Gerwig stated that her 2017 screenplay for *Lady Bird* was in part inspired by her personal and other people's memories of her teenage years; she even worked very closely with her cinematographer so that the film would "look like a memory."[24]

b. What happens when the source is a previous film (a feature, a short film, a documentary)—when, in other words, the film adaptation is based on a story found in the same medium?

c. What happens when the source is simply the screenplay, which cannot be considered another medium the same way a novel or a comic book can?

Given these questions, it is more helpful to state that *the object of film adaptation studies is the audiovisualization of written words or symbols/drawn images/ musical compositions/moving images/abstract images/mind images/objects in all their forms.*

This rationale leads to my argument that *all films are adaptations* of sorts (even those films that do not originate from screenplays adapt the verbal commands of the filmmakers and the thoughts of the actors regarding how to best "adapt" the story's basic theme or even lack thereof). The inclusive character of my proposition is extensive and does not prioritize one source over another. Still, we have to define what an adaptation is.

The question has already been addressed by several scholars, most of whom[25] begin by noticing that the word "adaptation" may indicate a process, a product, and its reception at the same time.[26] Although Corrigan[27] argues that an all-encompassing definition of "adaptation" is an impossible task—as most definitional attempts are in film studies—I would like to discuss three that pertain to the idea that all films are adaptations.

a.  Glenn Jellenik defines adaptation "as the frank appropriation of a text that taps into a new (and usually broader) audience or commercial market."[28]
b.  For Claus Clüver, adaptation "seems best used to cover the process (and its results) of adjusting a specific source text to the requirements and possibilities of another medium in such a way that parts of it are retained and incorporated in the resultant new text [medial configuration]."[29]
c.  Finally, Dennis Cutchins opts for a more general definition, as he finds that adaptation is "*a way of looking at texts.*"[30]

What these three definitions share is that, first, they aren't limited to adaptations that acknowledge their sources. Second, they are wide enough to accommodate all kinds of sources and their cinematic treatment. In 2017, Guillermo del Toro cowrote and directed *The Shape of Water*, based on a story of his own. *The Shape of Water* narrates the story of mute Elisa (Sally Hawkins), who works as a janitor at a secret and perhaps military US facility in the 1960s and inadvertently falls in love with a sea creature that is kept in a tank. As del Toro has shared in numerous interviews,[31] his story was inspired by the B sci-fi *Creature from the*

*Black Lagoon* (1954), which he saw when he was six or seven years old. Shot at the beginning of the Cold War and under the threat of the atomic bomb, the film follows a group of American scientists in their efforts to capture an amphibian sea monster in the Amazon, but meanwhile the creature develops feelings for the protagonist's female assistant. In del Toro's words, inspiration came when he first watched the film as a kid and especially the moment

> where the creature swims under Julie Adams in a white swimming suit. Three things awakened in me—one, Julie Adams. At six I was a horny little bastard. That's why this is my first movie that has full-on sexuality. The second that awakened was a Stendahl syndrome. There was something unassailable in that movie that I could not express. I got overwhelmed with beauty. And the third thing is, I felt a longing in my heart that I could not name. I kept thinking I hope they end up together and they didn't. So this is me correcting the cinematic mistake.[32]

This personal account underlines del Toro's investment in the project. In a way, he had been adapting *Creature from the Black Lagoon* ever since he first watched the film, but it took him thirty-five years to actually shoot his "adaptation" along with cowriter Vanessa Taylor. In addition, the filmmaker was also inspired by his environment, and his setting the film in the 1960s is not accidental. He remarks:

> The first thing is, I thought it was an ideal time to talk about love. The second thing is, when America talks about "Make America Great Again," it's talking about 1962, the end of Camelot, the peak of the promise of the future, jet-fin cars, super fast kitchens, television, everything that if you're white, Anglo-Saxon, heterosexual, you're good. But if you're anything else, you're not so good.[33]

Despite all this media attention on del Toro's film and his details regarding his sources of inspiration, controversy arose after *The Shape of Water* picked up thirteen Oscar nominations in January 2018. First, the estate of playwright Paul Zindel sued the director for plagiarizing Zindel's 1969 play, *Let Me Hear You*

*Whisper.*[34] Then, del Toro was accused of imitating the Dutch student short film *The Space Between Us* (2015), while French director Jean-Pierre Jeunet revealed that he was "troubled by a scene in *The Shape of Water* which bears striking similarities to a scene in his 1991 film *Delicatessen.*"[35] While the Dutch charges were dropped and the scene Jeunet refers to can only be considered an allusion, a visual quotation of sorts and not plagiarism, the dispute surrounding the "source(s)" of the film is certainly driven by the need for vindication, monetary compensation, and media attention on the part of the accusers, but it is also indicative of the polymorphia of adaptation in the twenty-first century. The compound noun "polymorphia" comes from the union of the Greek adjective πολύς (= many) and the noun μορφή (= form), signifying the quality of something or someone to adopt many forms. Thus, the polymorphia of adaptation may indicate the variety of sources that can inspire a new text or the use of "other" sources that were not popular or simply did not exist in the past (i.e., video games).

In the case of *The Shape of Water*, polymorphia signifies the former; Del Toro was inspired by a story he made up as a child, consciously expanding or adapting a film that had made a strong impression on him, which he then reworked with his cowriter, while also adapting the sociopolitical environment of the time of the film's production to comment on specific issues. His final product, a romance-centered, political, sci-fi, period film with rich intertextual layers—and, according to the credits and the Academy of Motion Pictures, based on an original screenplay—was accused of plagiarism, adapting without permission a play, a short film, and a feature film. Taking into consideration the myriad stories that exist in our time, it would be no exaggeration to suggest that almost all films that are based on original screenplays can be accused of plagiarism and also that *The Shape of Water* could have had many more accusers.

If we accept that every film is an adaptation of sorts, then adaptation studies becomes one *way of looking* at every film irrespective of its potential marketing/distribution as an adaptation of, let's say, a popular novel. It is up to the researcher to choose her *way of looking*. For instance, *The Shining* can be examined and analyzed as an adaptation of Stephen King's 1977 novel, but it can also be placed in an auteur analysis of Stanley Kubrick's oeuvre, an examination of Jack Nicholson's thespian evolution and/or stardom, or a theoretical and/or historical genre study of horror.

# The Theoretical and Methodological Conundrum

To this day, there is no unified, complete, or widely accepted theory to accompany adaptation studies despite its sixty-year existence (its birth is traditionally marked by Bluestone's book). What I mean by "theory" is not a series of insightful observations or a categorization of the different types of adaptation but a model of empirically selected and rather arbitrary elements that can be applied to the analysis of a wide number of objects and specifically, in the case of adaptation studies, texts of all shapes and sizes. Narrative theory originating from semiotics and structuralism may have provided tangible models for unlocking texts but has curiously ceased to be à la mode.

This theoretical void—which also features as the first misconception in Leitch's extensively cited article "Twelve Fallacies in Contemporary Adaptation Theory"[36]—is the result of five parameters. The first factor is the presumed death of grand narratives in the late 1970s and the postmodern attitude of deconstruction; the second is structuralism's failure to strongly articulate its theoretical models with contextual parameters that go beyond the text itself (i.e., the history of a text's production); the third regards academia and is twofold: on the one hand, there is a general avoidance of theorization in an era where Grand Theory had been declared dead and instead there arose a tendency toward adopting cultural and sociological approaches; on the other hand, it would be presumptuous to put forward a new textual theory or even a complementary one when Roland Barthes, Algirdas J. Greimas, and Gérard Genette, among others, have managed to leave us with narrative models whose application is widely disseminated and utilized even though not always acknowledged.

The fourth factor is the fact that adaptation studies, much like film studies, is in need not only of a textual theory or textual theories but also of an arsenal of theoretical models that can examine and interpret its object of study. For instance, distinguishing and comparing the operation of what Barthes calls the cardinal functions (main actions) or indices (elements that create the narrative's atmosphere) of a film with their operation in the texts it adapts is merely the first step in a case study examination between a film and its source. The questions that follow this initial process include queries, for example, Why were specific subplots and/or characters from a novel cut from or added to the final adapted text? Why has the adaptation's atmosphere changed from ominous to humorous? Why has the action moved from country X to country Y? Time restrictions imposed on cinema, production parameters, or even the nationality of a film-maker can provide some of the answers, but they barely explain the totality of omissions and/or changes encountered in an adaptation. Is it censorship? Is it a different approach chosen by the screenwriter/director/producer/star actor for specific reasons (aesthetic or otherwise)? Is it ideology? These last three questions point to another kind of methodology and research, as the scholar must either delve into archival material for the answer if the film belongs to the past or talk directly to the filmmakers if they can contact them. This work presumes not only the scholar's research abilities but also their knowledge of historiography principles, as the narrative that will ultimately be written will form part of adaptation history as well—not only of the single case study but of adaptation studies as a discipline.

These theoretical and methodological impasses, among others, have led adaptation scholars to complement their research with tools from sociology, psychoanalysis, gender studies, race studies, translation studies, and political science. After all, as Kamilla Elliott observes, "Adaptation studies is fractured because it participates in so many disciplines, periods, cultures, and media" and therefore borrows theories and methodologies from varied fields.[37] This brings me to Leitch's concluding remarks in his latest edited collection. After noting the historical character of theories, Leitch proposes the replacement of the single and still elusive Grand Theory with "petit theories." He asserts that this practice "can make us more receptive to several recommendations for the conduct of theory in general, and adaptation theory in particular."[38] Leitch supports both "an open-ended dialectic" of contrasting views and paradigms and a historical contextualization to avoid transforming scholarly work "into stifling dogmas."[39] Additionally, adaptation studies, and to be more specific, film adaptation studies, would greatly benefit from these practices:

a. historicizing its object (i.e., compiling corpora to assess the film adaptations during a specific time period in different countries and not just focusing on the American paradigm) and
b. revisiting the structural theories of the past, which can still reveal patterns and structures, but this time complementing this work with extratextual parameters to examine adaptations as sociohistorically constructed processes.

Adaptation is vital to disseminating important stories and myths as well as knowledge to new generations. In the 2018 Netflix sitcom *Champions*, Priya (Mindy Kaling) has to announce to Vince (Anders Holm), an aloof and rather irresponsible gym owner, that he has an openly gay teenage son he never knew about. Just before the revelation in his office at the gym, Vince spots teenage Michael (Josie Totah) there, and he initially thinks the kid is trying to steal something. When Michael tells him that he "was simply looking for something to eat, like Jean Valjean from *Les Miserables*," Vince replies that he has never read it. Michael then looks at him and dramatically declares: "It's not a book, it's a musical and a movie." I was very amused by this line until I asked my first-year students, who were born in 2000—around the time of fictional Michael's birth—if they knew whether James Whale's *Frankenstein* was based on an original screenplay. Much to my surprise, only 30 percent of the class knew it was an adaptation of a novel and only a much smaller percentage could identify the novel's author or her time. Yet, despite failing to answer a question that the greater part of Gen X would find easy, irrespective of their educational level, all my students still knew the monster of Frankenstein, its myth, and the basic struggles between man and God and between man and science that pervade the story. Just like fictional Michael, who was certain about Jean Valjean existing solely on Broadway and celluloid, my students believed, until corrected, that Frankenstein's monster was the result of a filmmaker's inspiration. That is when I started to grasp the immense potential of film adaptations in preserving (even by changing, adding, or subtracting from the "original") stories and myths that can still resonate decades or even centuries later. In this way, film adaptations become cultural gatekeepers, and as such their study is paramount.

## The Anthology

This collection is indicative of the plurality and polymorphia of adaptation studies' current historical period. As such, it is an eclectic and varied group of chapters, each of which focuses on a subject that helps adaptation studies move forward. The chapters, while generally heterogeneous, are divided into three parts.

Part I, "External Influences on Adaptation," delves into matters that encircle film adaptations without primarily focusing on textual analysis of the final cinematic product. The first chapter explores how American literature is adapted in foreign cinemas. In "Thirteen Ways of Screening American Literature," Thomas Leitch distinguishes thirteen different strategies that filmmakers around the world have used when confronted with an American literary work. This categorization is then used to discuss which strategies are more popular, the attitudes of such adaptations toward American myths, and what American literature as a whole means to the foreign cinematographies that use it as an adaptation source. My chapter, "The State of Contemporary Adaptation: A Revealing Corpus," aims at underlining the importance of data in adaptation studies, especially in the era of digital humanities, which afford us unlimited access to information. Although relevant contemporary publications still focus on paradigmatic case studies or methods of analysis and theories, the current adaptation landscape escapes us. It is important for adaptation scholars to know how many plays are adapted each year, if the adapted novels belong to the literary canon or are popular best sellers, the number and genre of comics preferred by filmmakers, and so on. These are some of the questions the chapter attempts to answer, based on a corpus that features all adapted films produced and/or coproduced in the United States (the cinematic paradigm par excellence) between 2006 and 2015. Finally, in "The Impact of Censorship on Adaptations: The Case of Portugal during the New State (1933–74)," Eurydice Da Silva examines how censorship, "a non-visible source," affects the process of adaptation and the final product. By unearthing archival material at the French Film Library in Paris, the Portuguese Film Library, and the National Archives of Torre do Tombo in Lisbon, Da Silva's impressive research uses the paradigm of censorship on film adaptations during the Portuguese New State (1933–74) to discuss how extratextual parameters are paramount in adaptation studies and how their study can even inform history.

Part II, "Millennial TV and Franchise Adaptations," demonstrates that the contemporary television landscape has become a very fruitful terrain for adaptation studies. The impressive numbers of popular long-form narratives that are based on a single sources or multiple ones complicate questions regarding the nature of the source or sources, the relationships between the "original" and the adapted text, the continuing force of culturally iconic characters and the trend of their "backstory," and the freedom with which showrunners and producers use the source, among others. The first two chapters of this section take on the questions above. After underlining how the area of television adaptations has not been theorized on any grand scale, Christina Wilkins, in "Television Adaptations: Character and World Expansion," explores what she labels "investigative adaptation," defined by the adaptation's need to function as explanatory and uncover the perceived whole of a character or world in the original text. Wilkins uses the shows *Hannibal, Bates Motel,* and *Westworld* as paradigms to flesh out how these adaptations offer an in-depth exploration of the workings behind the original text and its psychologically rich characters, giving the viewer a chance to see how the source text is constructed and can be understood. In this way, television adaptations provide the viewer with a blueprint with which to re-read and understand the original texts, and, with it, alter notions of the canon. Chapter 5 delves into TV horror and examines problematic cases of TV shows whose sources are either incomplete or have not reached an ending and televisual narratives that problematize notions of what exactly is adapted. Simon Brown and Stacey Abbott propose we view most TV horror adaptations as instances of inspiration from a source (irrespective of its nature or number) in "Inspiration as Adaptation: TV Horror, Seriality, and the Adapted Text." These TV narratives should be viewed as existing "somewhere between adaptation and inspiration, paying homage to the original text, but also establishing their own distinct televisual identity." In chapter 6, "Visuality, Continuity, and Coherence in Contemporary Fantasy Storyworlds," Karin Boklund-Lagopoulou takes on a different path as she examines three adaptation franchises from both cinema and television, focusing on the functions of visuality, an aspect of adaptation that is usually found in the margins of most analyses that emphasize the narrative and tend to take the visual cosmos created by filmmakers and television productions for granted. Boklund-Lagopoulou finds that visuality is a crucial element in creating coherence in audiovisual narratives and uses the narrative theory of Algirdas J. Greimas, and especially the concept of semantic isotopy, to analyze the visual style of her case studies. In this way, she achieves two things: first, she begins the discussion on the visual cosmos that accompanies the audiovisualization of a source (either written or audiovisual)—which should become part of film adaptation studies in general—and second, she shows how structuralism not only is still relevant but can also be applied to visuality.

Part III, "ElasTEXTity and Adaptation," contains the final four chapters of the collection. The term "elasTEXTity" was proposed by Grossman in her valuable monograph *Literature, Film, and Their Hideous Progeny: Adaptation and ElasTEXTity* and is defined as "a way of thinking about texts as extended beyond themselves, merging their identities with other works of art that follow and precede them . . . the state of being pulled beyond an initial form to encompass other objects, texts, and identities."[40] Although Grossman is also interested in the "monstrous results" this "elasTEXTity" can have, as the final adaptation can "appear to be misshapen or distorted,"[41] it is precisely the expanding factor of the term that is so useful for the freedom it allows the adapter to draw from the pool of myriad elements that consist in any given source and to create their own vision using the raw materials of their chosen medium if it happens to differ from the medium the source belongs to.

Chapter 7 shows one aspect of elasTEXTity, as it examines a Greek transnational adaptation of an American film, which, in its turn, originates in a classic tragedy. Ursula-Helen Kassaveti applies Arjun Appadurai's concept of mediascapes and Iain Robert Smith's model of proliferation/industry/reception to discuss this transnational adaptation in the context of the Greek musical genre of the 1980s. In "Transnational Adaptations in a Globalized Context: The Case of the Greek Film Musical," Kassaveti unearths the ways Greek filmmakers localize the predominantly American sources to draw in larger audiences and argues that it is only through a multiple factor study that transnational adaptations can reveal their wealth of meanings. Although elasTEXTity affords great freedom to artists, it can also lead to adaptations that pose difficult questions. Nicole Pizarro's chapter returns to the film/source paradigm to explore the implications of incorporating rape scenes as a narrative strategy in film adaptations of nonfictional slave narratives written by men, which actually do not include these violent assaults. Although Linda Hutcheon contends that adaptation "creates the doubled pleasure of the palimpsest,"[42] Pizarro argues that it also facilitates trauma through its transmedial nature. Pizarro also maintains that black women's bodies function as sites of cultural trauma and that just like the fictional women on screen have no way out, neither does the audience. The penultimate chapter 9, Joakim Hermansson's "Adaptations of Adulthood: Toward a Model for Thematic Rhetorics in Adaptation Studies," presents a model for the study of thematic arguments in adaptations through his investigation of the representation of adulthood in ten successful contemporary novel/film adaptations. Away from concentrating on the adaptation process or making evaluative comparisons between novels and films, Hermansson draws on Sarah Cardwell's definition of urtexts and Joseph Campbell's hero's journey and monomyth to examine how the thematic arguments and patterns develop in stories. With adulthood as a thematic example, the chapter demonstrates how each step of the structure invites specific questions and perspectives that can be applied to any thematic focus, as well as its rhetorical progression through a narrative. The author argues that this model opens up a new path for adaptation studies that will include corpora studies, with an impartial attitude toward the cinematic and literary versions of a story. The final chapter, Thomas Britt's "Experimental Music as a New Frontier of Adaptation Studies," exemplifies the argument regarding how the removal of the film/literature paradigm in adaptation studies can broaden and empower the field. Britt discusses a variety of musical case studies (a composer who creates a source and then adapts it for another purpose, musical creators who imagine a source that is unavailable to them, and a musical creator who adapts nonmusical sources into music) in order to "establish experimental music adaptation as a new frontier in adaptation studies because the creation and analysis of such works introduce some innovative approaches to production and distribution." Through his paradigm, Britt shows how musical adaptations "preserve, replicate, critique, repurpose, and/or imagine information from the past," extending the life of "former products and projects into the uncertain future." While this aspect of preservation is also a consequence of the film/source paradigm, Britt elevates this practice to another level by underlining the importance of this transportation of cultural memory.

Taken together, these ten chapters use wide-ranging approaches to contemporary adaptation, focusing on different perspectives and examining mainly adapted audiovisual texts through a variety of theoretical and methodological prisms. As such, they not only reflect the heterogeneity in which adaptation studies finds itself presently but also underline the importance of the field and its theories in the context of the current transmedia environment. The plurality of the collection allows the chapters to be read both separately and thematically, according to each reader's preoccupations and concerns. Thus, the reader with specific interests could dive immediately into relevant chapters regardless of order. Hopefully, this collection will not only complement existing literature but also raise new concerns and questions that can be taken up in the future. One thing remains certain: adaptation studies will continue to grow and evolve.

# Notes


1. Umberto Eco, *Reflections on "The Name of the Rose,"* trans. William Weaver (London: Minerva, 1994).

2. Thomas Leitch, "Introduction," in *The Oxford Handbook of Adaptation Studies,* ed. Thomas Leitch (Oxford: Oxford University Press, 2017), 1–22.

3. Robert Stam, *Literature through Film: Realism, Magic, and the Art of Adaptation* (London: Blackwell, 2005); and Robert Stam and Alessandra Raengo, eds., *A Companion to Literature and Film* (Malden, MA: Blackwell, 2004).

4. Leitch, "Introduction," 5.

5. Costas Constandinides, *From Film Adaptation to Post-Celluloid Adaptation* (New York: Continuum, 2010).

6. Kathleen Loock and Constantine Verevis, eds., *Film Remakes, Adaptations and Fan Productions* (New York: Palgrave, 2012).

7. Carolyn Jess-Cooke and Constantine Verevis, *Second Takes: Critical Approaches to the Film Sequel* (Albany: SUNY Press, 2012).

8. Julie Grossman, *Literature, Film, and Their Hideous Progeny: Adaptation and Elas-TEXTity* (New York: Palgrave, 2015).

9. Yvonne Griggs, *Adaptable TV* (New York: Palgrave, 2018).

10. Robyn McCallum, *Screen Adaptations and the Politics of Childhood* (New York: Palgrave, 2018).

11. Jeremy Strong, ed., *James Bond Uncovered* (New York: Palgrave, 2018).

12. Petr Bubeníček, *Subversive Adaptations: Czech Literature on Screen behind the Iron Curtain* (London: Palgrave Macmillan, 2017).

13. Iain Robert Smith, *The Hollywood Meme: Transnational Adaptations of World Cinema* (Edinburgh: Edinburgh University Press, 2017).

14. Dru Jeffries, *Comic Book Film Style: Cinema at 24 Panels per Second* (Austin: University of Texas Press, 2017).

15. Yvonne Griggs, *The Bloomsbury Introduction to Adaptation Studies: Adapting the Canon in Film, TV, Novels and Popular Culture* (New York: Bloomsbury, 2016).

16. Julie Grossman and R. Barton Palmer, eds., *Adaptation in Visual Culture: Images, Texts, and Their Multiple Worlds* (New York: Palgrave, 2017).

17. Douglas McFarland and Wesley King, eds., *John Huston as Adaptor* (Albany: SUNY Press, 2018).

18. Richard J. Hand and Andrew Purssell, *Adapting Graham Greene* (New York: Red Globe Press, 2015).

19. Wieland Schwanebeck and Douglas McFarland, eds., *Patricia Highsmith on Screen* (New York: Palgrave, 2018).

20. Max Sexton and Malcom Cook, *Adapting Science Fiction to Television* (Lanham, MD: Rowman & Littlefield, 2015).

21. To this series of monographs and collections, I am not including the steady flow of articles, several of which have already greatly contributed to the field, that have been published during the same time by long-standing and respected journals, such as *Literature/Film Quarterly* and *Adaptation*.

22. Thomas Leitch, "Adaptation and Intertextuality, or, What Isn't an Adaptation, and What Does It Matter?" in *A Companion to Literature, Film and Adaptation*, ed. Deborah Cartmell (Hoboken, NJ: Wiley-Blackwell, 2012), 91.

23. James Naremore, "Introduction: Film and the Reign of Adaptation," in *Film Adaptation*, ed. James Naremore (New Brunswick, NJ: Rutgers University Press, 2000), 15.

24. Julie Miller, "How Greta Gerwig's *Lady Bird* Came to 'Look Like a Memory,'" *Vanity Fair*, November 3, 2017, https://www.vanityfair.com/hollywood/2017/11/greta -gerwig-lady-bird-design.

25. See, for instance, Dudley Andrew, "Adaptation," in *Film Adaptation*, ed. James Naremore (New Brunswick, NJ: Rutgers University Press, 2000), 28–37; Timothy Corrigan, "Defining Adaptation," in Leitch, *Oxford Handbook of Adaptation Studies*, 23–35; and Claus Clüver, "Ekphrasis and Adaptation," in Leitch, *Oxford Handbook of Adaptation Studies*, 459–76.

26. Linda Hutcheon, *A Theory of Adaptation*, 2nd ed. (Oxon: Routledge, 2013), xvi.

27. Corrigan, "Defining Adaptation," 34.

28. Glenn Jellenik, "On the Origins of Adaptation, as Such: The Birth of a Simple Abstraction," in Leitch, *Oxford Handbook of Adaptation Studies*, 38.

29. Clüver, "Ekphrasis and Adaptation," 463–64.

30. Dennis Cutchins, "Bakhtin, Intertextuality, and Adaptation," in Leitch, *Oxford Handbook of Adaptation Studies*, 79.

31. See Rebecca Keegan, "How Guillermo del Toro Crafted the Perfect Monster-Romance For the Trump Era," *Vanity Fair*, September 4, 2017, https://www .vanityfair.com/hollywood/2017/09/guillermo-del-toro-shape-of-water; Borys Kit, "How Guillermo del Toro's 'Black Lagoon' Fantasy Inspired 'Shape of Water,'" *Hollywood Reporter*, November 3, 2017, https://www.hollywoodreporter.com/ news/how-guillermo-del-toros-black-lagoon-fantasy-inspired-shape-water -1053206; and Tim Gray, "Love and Danger on the 'Water' Front," *Variety*, January 10, 2018, https://variety.com/2018/film/awards/shape-of-water-inspiration -from-monster-movie-1202659976/.

32. Keegan, "How Guillermo del Toro Crafted."

33. Ibid.

34. Yohana Desta, "Guillermo del Toro Vigorously Defends *Shape of Water* against Plagiarism Claim," *Vanity Fair*, February 22, 2018, https://www.vanityfair.com/ hollywood/2018/02/guillermo-del-toro-defends-shape-of-water-lawsuit.

35. Eliza Berman, "Everything to Know about the *Shape of Water* Plagiarism Controversy," *Time*, March 3, 2018, http://time.com/5170613/shape-of-water-plagiarism -controversy/.

36. Thomas Leitch, "Twelve Fallacies in Contemporary Adaptation Theory," *Criticism* 45, no. 2 (2003): 149–71.

37. Kamilla Elliott, "How Do We Talk about Adaptation Studies Today?" *Literature/ Film Quarterly* 45, no. 2 (2017), https://lfq.salisbury.edu/_issues/first/how_do_we _talk_about_adaptation_studies_today.html.

38. Thomas Leitch, "Against Conclusions: Petit Theories and Adaptation Studies," in Leitch, *Oxford Handbook of Adaptation Studies*, 704.

39. Ibid., 705.

40. Grossman, *Literature, Film, and Their Hideous Progeny*, 2.

41. Ibid.

42. Hutcheon, *Theory of Adaptation*, 116.

# 1

# External Influences on Adaptation

# 1

# Thirteen Ways of Screening American Literature

## Thomas Leitch

WHEN *THE SALESMAN* WAS nominated for the 2017 Academy Award for Best Foreign Language Film, its director, Asghar Farhadi, announced that he would boycott the Oscars in protest against Donald Trump's executive order banning Iranians from entering the United States. The film's victory on Oscar night allowed Anousheh Ansari, who accepted the award on Farhadi's behalf, to read a statement Farhadi had prepared:

> I'm sorry I'm not with you tonight. My absence is out of respect for the people of my country and those of the six other nations who have been disrespected by the inhumane law that bans entry of immigrants to the U.S. Dividing the world into the us and our enemies categories creates fear, a deceitful justification for aggression and war. These wars prevent democracy and human rights in countries which themselves have been victims of aggression. Filmmakers can turn their cameras to capture shared human qualities and break stereotypes of various nationalities and religions. They create empathy between us and others—an empathy that we need today more than ever.[1]

Commentators around the world immediately seized on Farhadi's protest against what he saw as the oppression and inhumanity of the American ban on immigration on travelers from much of the Arab world, and his boycott became front-page news. But I'd like to draw attention to another aspect of his now widely known reaction to his Academy Award: his argument that films create an empathy that crosses national borders by capturing shared human qualities and breaking stereotypes.

Apart from the fact that it's just as easy for movies to create and exploit stereotypes as it is to break them—a phenomenon abundantly on display in the portrayals of German and especially Japanese nationals in Hollywood films produced during World War II—the most interesting feature of Farhadi's statement depends not on what it says but on its application to his own film. *The Salesman* stands out from most winners of the Academy Award for Best Foreign Language Film, and indeed from most non-American films of any sort, in being an adaptation, however free, of a well-known American literary property. *Death of a Salesman*, the 1949 Arthur Miller play in which Farhadi's two leading characters, Emad (Shahab Hosseini) and Rana (Taraneh Alidoosti), are acting, turns out to provide a wholly unexpected parallel to the characters' lives when a structural collapse in their apartment forces them to move into another apartment, whose previous tenant was a woman who departed recently and hastily. One night, a stranger enters the apartment while Rana is taking a bath, assaults her, and then leaves behind some money. When Emad tracks down the assailant, he turns out to be Majid (Mojtaba Pirzadeh), an elderly salesman and client of the prostitute who had formerly lived in the apartment. Locking Majid in a room in the apartment, Emad calls Majid's family to witness firsthand the revelation of his crime. Before they arrive, however, he realizes that Majid, agonizing over his impending shame, has suffered a heart attack. Instead of telling the story of the assault to Majid's family, who think Emad has summoned them to the scene of a medical emergency after saving Majid's life, Emad secretly slaps him, and Majid collapses as he is leaving with his family. After weeks of rehearsing the role of Miller's downtrodden salesman Willy Loman, Emad is forced to acknowledge that he has played an active role in humiliating and perhaps destroying the literal salesman in a parallel story in which he has turned out to play quite a different role than he would have expected.

Hollywood, as everyone knows, has always had a rapacious appetite for material from non-American cultures. The typical fate of foreign properties adapted to the American screen is colonial appropriation. The national identities and associations of the property's characters and their world are erased or attenuated, as in *Sommersby* (1993), a remake of Daniel Vigne's *Le retour de Martin Guerre* (1982), and *The Birdcage* (1996), a remake of Édouard Molinaro's *La cage aux folles* (1978), or reified and hypostasized in ways that emphasize, despite Farhadi's observation, their stereotypical foreignness, as in innumerable Hollywood versions of such foreign classics as Leo Tolstoy's *Anna Karenina* (1873–77) and Victor Hugo's *The Hunchback of Notre Dame* (1831). No wonder, then, that Jennifer L. Jeffers complains that "the 'Hollywoodization' of British literary texts"[2] amounts to a "colonization of Britain" by its former colony that enables "Hollywood's invention of Britain by adapting its literature"[3] or that Linda Hutcheon concludes that "for Hollywood . . . transculturating usually means Americanizing a work."[4] The work of American cultural imperialism that Jeffers traces through Hollywood appropriations of English literary sources over the twenty-year period from 1990 to 2010 finds a counterpart in its success in exporting the reterritorialized products. In a recent letter to the *New York Times Book Review*, for example, Jacqueline Chappel noted that "the market for translations from English far outpaces the market for translated works into English."[5]

It is all the more surprising, then, that *The Salesman* should represent a deviation from a pattern that is itself a glaring exception to the pattern of American cultural hegemony: the remarkable absence of adaptations of American novels, plays, and stories from foreign cinemas. This dearth was documented some forty years ago in Gerald Peary and Roger Shatzkin's collections *The Classic American Novel and the Movies*[6] and *The Modern American Novel and the Movies*,[7] whose combined eighty-five-page filmographies of adaptations of American novels include only twelve non-American films. The sixteen-page list of adaptations that concludes *The Classic American Novel and the Movies* includes just two foreign adaptations: *Dzhimmi Khiggins*, Georgi Tasin's lost 1928 Soviet adaptation of Upton Sinclair's 1919 novel *Jimmie Higgins*, and *Prinz und Bettelknabe*, Alexander Korda's 1920 German adaptation of Mark Twain's 1882 *The Prince and the Pauper*. The more extensive sixty-nine-page filmography of *The Modern American Novel and the Movies* adds ten more much better-known foreign adaptations of American properties: *Ossessione* (1942), based on James M. Cain's *The Postman Always Rings Twice* (1934); *Mr. Arkadin* (1955), based on Orson Welles's "Man of Mystery," a 1952 episode in the radio serial *The Adventures of*

*Harry Lime* (1951–52); *Purple Noon* (1960), based on Patricia Highsmith's *The Talented Mr. Ripley* (1955); *High and Low* (1963), based on Ed McBain's *King's Ransom* (1959); *Laughter in the Dark* (1969), based on Vladimir Nabokov's 1932 novel; *Ten Days' Wonder* (1971), based on Ellery Queen's 1948 novel; *Shoot the Piano Player* (1960), based on David Goodis's *Down There* (1956); *The Bride Wore Black* (1968), based on Cornell Woolrich's 1940 novel; *Mississippi Mermaid* (1969), based on William Irish's *Waltz into Darkness* (1947); and *Such a Gorgeous Kid Like Me* (1972), based on Henry Farrell's 1967 novel.

A great deal has changed in the forty years since Peary and Shatzkin did their head count. But one thing that has remained is the relatively small number of foreign films that adapt American literary properties. Of the thirteen essays Iain Robert Smith and Constantine Verevis collect in *Transnational Film Remakes*,[8] only two—Kenneth Chan's examination of Yimou Zhang's *A Woman, a Gun and a Noodle Shop* (2009)[9] and Rashna Wadia Richards's study of Sanjay Gupta's *Kaante* (2002)[10]—deal with remakes of American films, and in both of these cases the earlier films, *Blood Simple* (1984) and *Reservoir Dogs* (1992), are based on original screenplays rather than works of American literature. Part of the reason for this neglect is clearly the United States' tendency toward wholesale cultural assimilation in the name of what Jeffers has identified as a deliberately constructed national universalism that seeks to identify America with global values.[11] The United States has such a long history of having welcomed émigrés from around the world—a history that made it all the more inviting for Farhadi to call out the extraordinary restrictions on immigrants from Arab countries—that it has hardened into a cultural ideology of the melting pot, or the salad bowl, or the patchwork quilt: three metaphors that emphasize different aspects of hyphenate-Americanism while remaining constant in their emphasis on the American national project as assimilationist to its very foundations. Hollywood itself has been periodically renewed by the arrival of émigré screenwriters and directors, such as Fritz Lang, Billy Wilder, Jean Renoir, Robert Siodmak, and Douglas Sirk. Many of these filmmakers, even after settling in America, have maintained a continued interest in the novels and plays of their native lands. During Alfred Hitchcock's first decade in Hollywood, for example, all his adaptations—*Rebecca* (1940), based on Daphne du Maurier's 1938 novel; *Suspicion* (1941), based on Francis Iles's *Before the Fact* (1932); *Spellbound* (1945), based on Francis Beeding's *The House of Dr. Edwardes* (1928); *The Paradine Case* (1947), based on Robert Hichens's 1933 novel; *Rope* (1948), based on Patrick

Hamilton's 1929 play; and *Stage Fright* (1950), based on Selwyn Jepson's *Man Running* (1947)—were based on English novels except for *Under Capricorn* (1949), which was based on the 1937 novel by the Australian Helen Simpson. The locales of *Spellbound* and *Rope* were changed from England to America, and Hitchcock would later set both *Vertigo* (1958), based on Pierre Boileau and Thomas Narcejac's French novel *D'entre les morts* (1954), and *Family Plot* (1976), based on Victor Canning's English novel *The Rainbird Pattern* (1972), in San Francisco. The rest of these films retained their English settings; *Strangers on a Train* (1951) marked the first time that Hitchcock adapted an American source for a film set in America since *The Mountain Eagle* (1927). Although he did not display anything like Hitchcock's Anglophile preservationism, Ernst Lubitsch still continued to the end of his career to adapt European plays into such apparently all-American films as *The Shop Around the Corner* (1940), *That Uncertain Feeling* (1941), and *Heaven Can Wait* (1943). For whatever reason, Lubitsch's and Hitchcock's experiences as expert importers and rebranders of foreign properties find no parallel in stories of American filmmakers who have sought their fortunes abroad by exporting American literature.

As Farhadi points out, foreign filmmakers who do turn their attention to American literary properties may well seek to capture shared human qualities and break stereotypes of various nationalities and religions. A brief survey of foreign adaptations of American literature, however, indicates that the truth is more complicated than that, not because foreign filmmakers have other goals but because there are so many ways to pursue these goals, some of them quite at odds with each other. Concluding her analysis of *Ossessione*, Visconti's influential 1942 adaptation of James M. Cain's 1934 noir romance *The Postman Always Rings Twice*, Cristina Della Coletta refers to "Visconti's dialogue with American literature."[12] The phrase is telling because it implies that in addition to adapting a single novel to a single film, Visconti is engaging more broadly with the corpus and tropism of American literature as such. Following Hutcheon, who has ruled that we read differently if we focus on "an adaptation *as an adaptation*,"[13] I'd like to consider a range of foreign adaptations of American literature as adaptations *of American literature*, with special attention to the different strategies they adopt toward either America as such or American literature as such. In the process, I borrow and complicate the three models of national cinema that Darrell William Davis[14] proposed: a "reflectionist" model in which "film reflects preexisting cultures," so that "French, English, or Japanese film is basically the way it is because that is how French, English, or Japanese people and their society are"; a "dialogic" model that poses non-American cinemas as more or less challenging alternatives to Hollywood practice, so that "to the question of why Japanese cinema is special, the answer is because it relates in arresting ways to Western cinema";[15] and a "contamination" model that "avoid[s] binary categories like black-white, east-west, or *uchi-soto* (inside-outside, a distinction beloved by cultural anthropologists working on Japan)" to define national cultures as "fabricated piecemeal out of available bits and fragments, often from outside national borders,"[16] so that "national specificities jostle, catalyze, and 'thicken' without eclipsing or canceling one another out and without synthesizing into some new postnational order."[17] Unlike Davis's three models, which are designed to be mutually exclusive, my thirteen strategies inevitably overlap both logically (some of them assume the existence of others they build on or react against) and pragmatically (many adaptations incorporate more than one of these strategies). Although I regret the overlap that prevents them from being tidier, I make no apology for the possibility that highly reified, programmatic notions of America and American literature may emerge from my analysis of thirteen different attitudes that foreign adaptations have taken toward these subjects, for I'd argue that these notions are baked into the adaptations and only released, not imposed, by my analysis.

1. The most readily predictable of these strategies is the straightforward replication of American franchises by exporting their leading figures, as foreign superhero movies have done for many years, often without any explicit acknowledgment of their source material in their production credits or any payment to the producers of that material. *The Batwoman* (1968) presents a wealthy Mexican who uses her martial arts skills to bring justice to her city. *3 Dev Adam* (1973), or *3 Giant Men*, pits Captain America and Mexican wrestler Santo against the Istanbul criminal gang led by Spider-Man. A Japanese Spider-Man drives a Spider-Man car and shoots villains with a machine gun in the thirty-eight episodes of the 1978 television series *Supaidaman*. The hallmark of these adaptations, like that of Hollywood superhero sequels, is the invention of new but generically recognizable adventures for iconically costumed characters. These adaptations indigenize the problems superheroes must solve and the territories within which they must solve them even as they retain recognizably American heroes who happen to speak the language of their target audience abroad. Yet even this apparently straightforward appropriation can be complicated by a

meta-consciousness of the heroes' role-playing as role-playing. In *3 Dev Adam*, as Iain Robert Smith notes, Captain America displays no superpowers because he is dressing up as a superhero to catch "'a child-minded lunatic' who attacks anyone else who wears a mask. . . . [T]he use of superhero costumes is explained rationally as a method for crime prevention—in other words, we are not watching 'Captain America' in Turkey but rather a police chief who is dressed in that costume in order to capture a criminal."[18]

2. A more complex attitude that grows out of this drive to colonize the colonizer appears in foreign films that subject the colonizing impulse of American culture to more critical analysis. *Der amerikanischer Freund*, the German 1977 adaptation of Patricia Highsmith's 1974 novel *Ripley's Game*, focuses not on its title character but on Jonathan Zimmermann (Bruno Ganz), a mortally ill Swiss art forger who is manipulated into an impossible situation by the sinister Ripley (Dennis Hopper), an American whose shadow, even during the long periods between his brief appearances, is more powerful than the agency of Zimmerman or indeed of any of the other characters with whom he interacts. In contrast, the Italian/UK/US coproduction *Ripley's Game* (2002) gives its Ripley (John Malkovich) as much screen time and a good deal more prominence in its action sequences than his dying agent Jonathan Trevanny (Dougray Scott). The UK/US/Indian coproduction *Bride and Prejudice* (2004) not only turns Jane Austen's story of salt-and-pepper courtship into a cross-cultural romance between East and West but turns the very English Mr. Darcy into the very American Will Darcy. This change allows the heroine to denounce the colonizing impetus not of her homeland's historical Western adversaries, the English who established the Raj, but of the Americans, whose global capitalism is the greater threat to her native culture in 2004, before she and Darcy finally yield to each other in what Chadha presents as a compromise that is ideological as well as romantic.

3. Instead of being complicated by critique, the colonizing impulse of American culture can simply be reversed, as it is in the 1968 anthology film *Histoires extraordinaires* (*Spirits of the Dead*), which uses three stories by Edgar Allan Poe as the basis for three films in which European cultural concerns are played out against European cultural settings. It cannot be fairly said that the three films evacuate the American content of Poe's stories, since Poe sets one of them, "William Wilson," in England and a second, "Metzengerstein," in central Europe. Roger Vadim casts his American wife, Jane Fonda, as the Contessa Frederique de Metzengerstein and Peter Fonda as her cousin. But the other two

films push Poe's de-Americanizing tendencies still further. Louis Malle changes the setting of "William Wilson" from England to France, and Federico Fellini uses the all-American "Never Bet the Devil Your Head" as a vehicle for Terence Stamp, who stars as Toby Dammit, an unforgettably decadent British actor called to Rome, where he engages in every conceivable sort of transgressive behavior before the death broadly hinted in Poe's title.

4. Pursuing this last tendency back to the American homeland, filmmakers can cast America as an exotic place like the Land of Oz, where marvelous adventures are the order of the day. This attitude follows Philip Rahv in seeing American culture as polarized "between energy and sensibility, between conduct and theories of conduct, between life conceived as an opportunity and life conceived as a discipline."[19] Writing in 1939, Rahv famously parceled out American literature between palefaces—highbrow mandarins like Ralph Waldo Emerson, Nathaniel Hawthorne, and Henry James—and redskins—lowbrow entertainers like Walt Whitman, Mark Twain, and twentieth-century novelists like Theodore Dreiser, Sherwood Anderson, and Sinclair Lewis—and announced that "at present the redskins are in command of the literary situation."[20] Foreign cinema's embrace of American redskins, emerging most clearly in the *Lederstrumpf* films, German silent films based on J. Fenimore Cooper's *Leatherstocking Tales*, persists in more recent Cooper adaptations like the East German television segment *Der letzte die Mohikaner* (1956) and feature film *Chingachgook, die grosse Schlange* (1967). These Cold War adaptations of the *Leatherstocking Tales* use the tight-knit communities of Cooper's Mohicans to sharpen redskin American writers' critique of American culture in the name of spontaneity, individuality, and experience by celebrating specifically socialist values they find missing from modern America.

5. Foreign adaptations can use their culturally marginal positions to contest First World cultural hegemony by talking back to America in the name of a transcultural commons. Robert Stam, the leading proponent of this position, has examined it mostly in Third World adaptations of English and Continental classics like Ketan Mehta's 1993 *Maya Memsaab*, an Indian retelling of *Madame Bovary*; Joseph Gai Ramaka's 2002 *Karmen Gei*, a queer Senegalese retelling of *Carmen*; and Lúcia Murat's 2008 *Maré, Nossa História de Amor*, a Brazilian *Romeo and Juliet*. More recently, Petr Bubeníček has analyzed a series of Cold War adaptations in which Czech filmmakers adapted similar strategies

to talk back to the Communist regime ruling their own country in films from *Silvery Wind* (1954) to *The Joke* (1969).[21] Attempts to work the same changes on American literature are harder to find, but there have been sporadic attempts to relocate American properties in a transcultural commons, from the 1969 Brazilian television series *A Cabana do Pai Tomás* (*Uncle Tom's Cabin*) to *El hombre largo*, a 2004 adaptation of Poe's story "The Oblong Box" for the Argentinian television series *Cuentos clásicos de terror*.

6. Some acts of cultural appropriation can express more complex, contradictory, or problematic attitudes. Stam and Joao Luiz Viera have argued, for example, that "parody is especially well suited to the needs of the oppressed and the powerless, precisely because it *assumes* the force of the dominant discourse, only to deploy that force, through a kind of artistic jujitsu, *against* domination."[22] Smith, dissenting from Viera and Stam's jujitsu metaphor as unnecessarily dualistic and reductive, acknowledges that their parodic model "points to the more ambivalent nature of transnational adaptations of Hollywood in general."[23] *Bacalhau* (1975) presents a Brazilian beach threatened by a monstrous cod from Guinea that can be defeated only by a courageous fisherman and a Portuguese oceanographer using records made by iconic Brazilian fado singer Amália Rodrigues, Portugal's best-selling recording artist, as bait. Although the film's credits never mention either Peter Benchley's 1973 novel or Steven Spielberg's 1975 film, its parody of *Jaws* is signaled by innumerable narrative and visual elements, beginning with its advertising poster. Yet the targets of its satire are largely indigenized. Instead of simply making fun of the Hollywood blockbuster, *Bacalhau* uses the leading elements of that blockbuster to make fun of targets in its own country as well. *Alyas Batman en Robin* (1991) is a Philippine parody that features ordinary heroes who, like the police chief who dresses as Captain America in *3 Dev Adam*, don the costumes Adam West and Burt Ward wore in the American television series *Batman* (ABC, 1966–68) to catch criminals who have disguised themselves as the Penguin and the Joker; songs sending up American pop tunes of the sixties; bank robberies punctuated by musical numbers; and a finale in which superheroes from Marvel and DC dance together as they sing, "Let's be afraid of God! Let's believe in love!" Perhaps strangest of all is Philippe Mora's 1983 Australian film, *The Return of Captain Invincible*, a musical parody of *Captain America* whose eponymous American hero flees to Australia after he is accused of being a Communist and

prosecuted for wearing underwear in public. Captain Invincible is persuaded to don his forbidden costume once more only when his archenemy, Captain Midnight, threatens the United States in ways that require a superhero's defense. The hero's transnational odyssey, which provided a metonym for the film's own intertextual roots, generated an unexpected debate after the film's release when Tom McVeigh, Australia's minister of home affairs, ruled the film "not Australian" and therefore ineligible for the 150 percent tax deduction due its investors. Producer Andrew Gaty sued to establish the film's Australian status and won the resulting court proceeding, which was more successful in raising questions about what exactly established a film's national character than in resolving the question of Captain Invincible's citizenship.

7. Instead of adopting complex or contradictory attitudes toward America, some adaptations seek simply to erase the Americanism of the texts they adapt. For instance, *Masoom* (1983) Indianizes all the characters of Erich Segal's 1980 novel *Man, Woman and Child*, which had already been adapted for American audiences by Dick Richards's 1983 film, the source of Kapoor's film. More recently, Harlan Coben's 2001 thriller *Tell No One* was adapted to a French setting in *Ne le dis à personne* (2006) after Coben's Hollywood option lapsed. Non-American adventures of *Lolita* from the French *Emanuelle e Lolita* (1978) to the Japanese *Itazura Lolita: Ushirokara virgin* (1986) borrow nothing from Vladimir Nabokov's 1955 novel but the promise of sex with a Lolita no longer identified as American but simply presented as a sex toy. In adopting this strategy, these foreign films eliminate one side of Nabokov's cross-cultural anatomy, the émigré professor Humbert Humbert's confrontation with a postwar America marked by endless highways and functional motels, in favor of the other side, which explores the pleasures and costs of sex between an adult male and an underage girl.

8. Alternatively, foreign filmmakers can follow the example of films about the Land of Oz by deculturating the American antecedents of the properties they are adapting, not in an effort to renationalize them but in search of an elemental primitivism or a transcendent universalism. L. Frank Baum's Oz books, whose film versions were limited to Hollywood until 1960, have since spawned television episodes, many of them animated—in Spain ("El mago de Oz," a 1964 episode of *Nuestro amigo el libro*), Finland (*Trollkarlen från Oz*, 1965), England ("The Wizard of Oz," in *Jackanory*, 1970), Japan (*Ozu no mahôtsukai*, 1970; the

fifty-two episodes of *Ozu no mahôtsukai*, 1986–87; the twenty-six episodes of *Supêsu Ozu no bôken*, 1992), Mexico (*Mago de Oz Cuento de Frank Baum*, 1985), Lithuania (*Geltonu plytu kelias*, 1993), and Russia (*Volshevnik Izumrudnogo goroda*, 1995, and *Priklyucheniya v Izumrudnom Gorode*, 1999–2000)—and feature-length films and videos—in Turkey (*Aysecik ve sihirli cüceler rüyalar ülkesinde*, 1971), Brazil (*Os Trapalhões e o Mágico de Oróz*, 1984), Canada (*The Wonderful Wizard of Oz, The Marvelous Land of Oz, Ozma of Oz,* and *The Emerald City of Oz*, all 1987), and Israel (*Hakosem!*, 1994, and *Ha-Kosem Me'eretz Utz Hamachzemer*, 2015). The globalization of the Oz franchise represents the ease with which Baum's distinctively American fairy tales can be universalized by deemphasizing or deculturating the avowedly American values of charmingly preadolescent female independence, republican government, and a do-it-yourself fondness for technological invention. Three of the four foreign film adaptations of *The Postman Always Rings Twice* use Cain's tale of love and murder to comment on contradictions particular to their own cultures, but Fehér György's 1997 Hungarian film *Szenvedély* comes closest to this position.

9. In addition to adapting specific, identifiable literary texts from the American canon, foreign adaptations may choose to focus on embedded figurations of either America or the adaptive process. Della Coletta argues that foregrounding and insisting on the "self-conscious kitsch" that marks the climax of "Toby Dammit," Fellini's segment from the *Spirits of the Dead*, make audience members "active contributors to Fellini's parodic creative process" even as it provides "an allegory of Fellini's relationship to Poe and of his 'free' adaptation of 'Never Bet the Devil Your Head.'"[24] Several episodes in the 1974–75 London Weekend Television program *Affairs of the Heart*, a series of fifty-minute adaptations based on Henry James's fiction, trace the transatlantic romances of *The Wings of the Dove*, "An International Episode," and "Daisy Miller." The transcultural vicissitudes that doom these romances in James provide suggestive figurations of both the Americanism of their heroines and the struggles to maintain one's identity while adapting to the mores of a foreign culture even as the adaptations reframe them within a specifically English context. "Bessie" (1974), for example, follows "An International Episode" closely after adding a brief new opening scene that roots the story of Lord Lambeth's visit to America firmly in England, beginning with a close-up of his mother, the marchioness, impatiently telling her son, "I cannot conceive *why* you wish to visit America anyway." The

effect is to de-Americanize the American perspective James brought to "An International Episode," a comic inversion of "Daisy Miller" written only a few years after James had settled in London.

10. The rationale for any number of foreign adaptations is the search for more generally illuminating analogies between adapted and adapting cultural moments, events, and contexts. Unlike *Szevnvedély*, the two earlier European adaptations of *The Postman Always Rings Twice*—*Ossessione* and its 1939 French predecessor, *Le dernier tournant*—both transpose Cain's critique of California culture, whose failure to deliver on its Edenic promise of new beginnings drives Cain's lovers to adultery, murder, and despair, to analogous contradictions in their own very different cultures. So does the 2008 German film *Jerichow*, which, without citing Cain in its credits, recasts his Greek American cuckold Nick Papadakis as Turkish German businessman Ali Özkan in order to explore deeper rifts between native-born Germans and recent Middle Eastern immigrants. Smith has traced the power of the Turkish horror film *Şeytan* (1974) to its Islamization of the Catholicism in *The Exorcist* (1973), based on William Peter Blatty's 1971 novel:

> Both *The Exorcist* and *Şeytan* deal with the perceived failings of scientific rationalism and propose the continued relevance of religion in the modern world. . . . [S]uch an argument had special resonance in Turkey in the mid-1970s, with *Şeytan* produced at a time when the Kemalist embrace of Western conceptions of modernisation and rationalism was again coming to the fore in Turkish political life.[25]

Hence Erksan's adaptation "allowed for a subtle form of political commentary on the status of Islam within Turkey which may have run into trouble had it been approached directly"[26] instead of under the aegis of a well-known American film.

11. Foreign adaptations can reveal the power of the American authors as minority voices against majoritarian American culture even if those authors' works are not directly identified as sources of the adaptation at hand. Setting the vagabond in *Ossessione* against the American writer Jack London's self-portrayal as "the quintessential American hobo" who "provided an estranging perspective on middle-class standards and mores and became the positive pole

against which the hypocrisy, corruption, and vulgarity of the moneyed classes shone forth in grotesque and lurid glory," Della Coletta argues that Visconti's antihero, unlike the more successfully rebellious London, "ends up confirming, rather than subverting, the power of the system that he only superficially appears to challenge."[27] Even so, the import of *Ossessione* is not "unredeemable circularity and solipsistic pessimism" but a "comparison between the corporatist and authoritarian ethos of the fascist regime and the founding tenets of American democracy: self-reliance, individualism, and personal accountability."[28] Commentators who reject this reading must still acknowledge the force of Della Coletta's distinction between American literature and American culture.

12. Although many adaptations combine tactics or motifs from several of the strategies I've been describing, some of them make a special point of pressing obviously contradictory strategies in variously hybrid, palimpsestuous, or carnivalesque forms. *Shoot the Piano Player* (1960), an adaptation whose gangsters, unlike their literary models, manage to be both menacing and clueless and whose hero is both affecting and affectless, is at once a valentine to noir fiction and Hollywood genres and a declaration of independence from precisely those modes. The dizzying array of cultural referents, modes, and attitudes in *The Great Gatsby* (2013), an Australian/US coproduction released by Warner Bros., positions this adaptation as both inside and outside F. Scott Fitzgerald's novel, its Jazz Age background, and American culture generally.

13. Several of the adaptations I have mentioned—*3 Dev Adam, Jerichow,* and, most famously, *Ossessione*—are unacknowledged adaptations that cross between America and foreign countries by sneaking over the borders. So are the great majority of Bollywood adaptations and remakes. Anupama Chandra has observed that "ninety percent of the Hindi movies in production in August of 1993 were remakes, either of earlier Hindi films, Hong Kong movies or Hollywood blockbusters."[29] It is inevitable that a certain number of Hindi films would remake movies based in turn on American literary properties. *Ek Ruka Hia Faisla* (1986) is a remake of *12 Angry Men* (1957), based on Reginald Rose's 1954 play; *Baazigar* (1993) is a remake of *A Kiss Before Dying* (1956), based on Ira Levin's 1953 novel; *Sangharsh* (1999) is a remake of *The Silence of the Lambs* (1991), based on Thomas Harris's 1988 novel; and *Black* (2005) is a remake of *The Miracle Worker* (1962), based on William Gibson's 1957 TV script and his 1959 play. Smith prefaces his discussion of *Sarkar* (2005), a remake of *The Godfather* (1972), by listing eight other films that "have taken inspiration from

Francis Ford Coppola's adaptation of the Mario Puzo novel . . . *Dharmatma* (1975), *Aadalat* (1976), *Nayakan* (1987), *Zulm ki Hakumat* (1992), *Aatank hi Aatank* (1995), *Vishwasghaat* (1996), *Mumbai Godfather* (2005), and *Family* (2005)."[30] What sets *Sarkar* apart from these earlier films, Smith observes, is that it "is explicitly positioned as an adaptation, with the director acknowledging the influence of the film in a director's note which opens the film: 'Like countless directors all over the world, I have been deeply influenced by *The Godfather*. *Sarkar* is my tribute to it.'"[31] Even here, however, Varma credits Coppola's film without mentioning the 1969 novel on which it is based. Lucia Krämer has noted that formally acknowledged adaptations whose credits list their literary sources are the exception rather than the rule in the Indian film industry, which has normalized the practice of "unacknowledged transcultural remaking."[32] Smith introduces his monograph on Turkish, Philippine, and Indian remakes by noting that "the films I am dealing with do not license their cross-cultural borrowings."[33] Patrick Cattrysse, going still further, suggests that "pseudo-originals or hidden/secret adaptations greatly number overt adaptations"[34] around the world. Adaptations produced under these assumptions seek to erase not so much any traces of American culture in their stories but any legally actionable indications of their sources as such. The potentially limitless number of these adaptations makes it hard to determine how representative any brief survey like this one may be of the entire field.

This list of thirteen stances that foreign adaptations have taken to American literary properties and the strategies they have used to adapt them is not meant to be exhaustive. Naghmeh Rezaie's Cross Cultural Adaptations website[35] makes it clear that *The Salesman* is far from the only Iranian film to take its inspiration from American literature by listing several other such films: *Captain Khorshid* (1987), based on Ernest Hemingway's 1937 novel *To Have and Have Not*; *Hamoon* (1990), based on Saul Bellow's *Herzog* (1964); *Pari* (1995), based on J. D. Salinger's *Franny and Zooey* (1961); and *Here without Me* (2011), based on Tennessee Williams's *The Glass Menagerie* (1944). Rezaie's dissertation research has expanded this list further, identifying *Topoli* (1972) as an adaptation of John Steinbeck's *Of Mice and Men* (1937), *The Storm* (1980) as an adaptation of Steinbeck's *The Pearl* (1947), *The Leaf and Wind* (1984) as an adaptation of O. Henry's story "The Last Leaf" (1907), *The End of Childhood* (1993) as an adaptation of Howard Fast's *The Children* (1947), *The*

*Stranger* (2011) as an adaptation of Williams's *A Streetcar Named Desire* (1947), and both *The Window* (1970) and *Leaning on the Wind* (2000) as adaptations of Dreiser's *An American Tragedy* (1925). I have not seen any of these films and doubt that most non-Iranian readers have done so either. Clearly, there is much more work to be done on foreign adaptations of American literature.

Instead of considering other possible stances, however, I'd like to conclude by highlighting an intention conspicuous by its glaring absence from the foreign adaptations of American literature I have seen: the replication or repurposing of specifically American myths about America. For all the interest East German cinema took in pressing into service the outsider's perspective of the *Leatherstocking Tales* on American social self-mythologizing, foreign adaptations have been pointedly uninterested in either recycling or deconstructing self-defining American myths like the vital importance of the American frontier, the distinctive status of the United States as a melting pot or patchwork quilt of immigrants or an arena for the American dream of individual or generational economic success, or the exceptionalism of American culture or its alleged congruence with Western or universal culture as such. Whatever foreign non-American filmmakers are using American literature to do, they persist in going their own way, which may well be the single most mythopoetically American feature of their work.

My abbreviated survey suggests several conclusions. One is the observation that even foreign filmmakers who are steeped in reverence for Hollywood films are much less likely to find cultural capital in the literary properties on which many of these films are based. American cinema, for better or worse, is a hotter export commodity than American literature. When non-American filmmakers do adapt American literary properties, particularly in India, home to the world's largest national cinema, they are more likely to approach them as remakes of Hollywood films than as readaptations of the literary properties behind those films and very likely to pass over their sources, literary or cinematic, without explicit acknowledgment. More fundamentally, it seems clear that whatever opinions non-American filmmakers may hold as private citizens, their adaptations of American properties are more invested in American cinema than American culture and more invested in American culture, especially American popular culture, than either American self-mythologizing or American literature. American cinema may continue to set standards for national cinemas around the world, but American literature is most often found on foreign screens when it has sneaked over the border and is hiding, however it may be disguised, in plain sight.

It is hardly surprising that of Davis's three models for national cinemas, transnational adaptations of American literature are far more consistent with the contamination model than either the reflectionist or the dialogic models. Part of the reason is self-evident: any non-American filmmaker seeking to adapt a work of American literature to a foreign screen is by definition courting a kind of contamination that challenges both the reflectionist (Japanese cinema reflects the Japanese culture with which it is seamlessly continuous) and dialogic (Japanese cinema is best defined as an exception to Hollywood practice) models. It is equally true, however, that the many different attitudes foreign filmmakers have taken toward American literary originals show that both American cinema and culture and their own national cinemas and cultures are already palimpsests rather than networks, defined partly by insiders, partly by outsiders, and partly by liminal figures and practices, in incessantly changing ways. One constant, however, is the rifts they continue to reveal between the nation's cinema, literature, and culture, which are themselves always already contaminated with essences, opposites, misfit elements, and wild cards of every kind. American literature, to take the most obvious case, can hardly be assumed to be congruent with majoritarian American culture, since American novels from *The Scarlet Letter, Moby Dick*, and *The Adventures of Huckleberry Finn* to *Lolita*, Philip Roth's *The Plot Against America*, and Toni Morrison's *Beloved* have repeatedly and explicitly taken it upon themselves to challenge, subvert, or deconstruct that culture—or, more precisely, what they take to be consensual myths about that culture. As Rahv's distinction between palefaces and Indians reminds us, American literature is a concept no more homogeneous or unified than American culture, and foreign filmmakers have presumably avoided it as a category for the same reason that so many American filmmakers have.

## Notes

1. Steve Dove, "Asghar Farhadi Oscar 2017 Winner Speech Delivered by Anousheh Ansari," *Oscar.go.com*, February 27, 2017, http://oscar.go.com/news/ winners/asghar-farhadi-oscar-2017-winner-speech-delivered-by-anousheh -ansari.

2. Jennifer L. Jeffers, *Britain Colonized: Hollywood's Appropriation of British Literature* (New York: Palgrave Macmillan, 2006), 5.

3. Ibid., 3.

4. Linda Hutcheon, *A Theory of Adaptation*, 2nd ed. (Oxon: Routledge, 2013), 146.

5. Jacquelyn Chappel, "Lost in Translation," *New York Times Book Review*, September 11, 2016, https://www.nytimes.com/2016/09/11/books/review/letters-to-the-editor.html.

6. Gerald Peary and Roger Shatzkin, eds., *The Classic American Novel and the Movies* (New York: Ungar, 1977).

7. Gerald Peary and Roger Shatzkin, eds., *The Modern American Novel and the Movies* (New York: Ungar, 1978).

8. Iain Robert Smith and Constantine Verevis, eds., *Transnational Film Remakes* (Edinburgh: Edinburgh University Press, 2017).

9. Kenneth Chan, "The Chinese Cinematic Remake as Transnational Appeal: Zhang Yimou's *A Woman, a Gun and a Noodle Shop*," in *Transnational Film Remakes*, ed. Iain Robert Smith and Constantine Verevis (Edinburgh: Edinburgh University Press, 2017), 87–102.

10. Washna Radia Richards, "Translating Cool: Cinematic Exchange between Hong Kong, Hollywood and Bollywood," in Smith and Verevis, *Transnational Film Remakes*, 118–29.

11. Jeffers, *Britain Colonized*.

12. Cristina Della Coletta, *When Stories Travel: Cross-Cultural Encounters between Fiction and Film* (Baltimore: Johns Hopkins University Press, 2012), 66.

13. Hutcheon, *Theory of Adaptation*, 6.

14. Darrell William Davis, "Reigniting Japanese Tradition with *Hana-Bi*," *Cinema Journal*, 40, no. 4 (2001): 62.

15. Ibid., 63.

16. Ibid., 65.

17. Ibid., 65–66.

18. Iain Robert Smith, *The Hollywood Meme: Transnational Adaptations of World Cinema* (Edinburgh: Edinburgh University Press, 2017), 48.

19. Philip Rahv, "Paleface and Redskin," *Kenyon Review* 1, no. 3 (1939): 251.

20. Ibid., 254.

21. Petr Bubeníček, *Subversive Adaptations: Czech Literature on Screen Behind the Iron Curtain* (London: Palgrave Macmillan, 2017).

22. Joao Luiz Viera and Robert Stam, "Parody and Marginality: The Case of Brazilian Cinema," *Framework* 28, special issue (1985): 22.

23. Smith, *Hollywood Meme*, 27.

24. Della Coletta, *When Stories Travel*, 105.

25. Smith, *Hollywood Meme*, 67.

26. Ibid., 65.

27. Della Coletta, *When Stories Travel*, 59.

28. Ibid., 66.

29. Anupama Chandra, "Cool Copy Cats," *India Today*, August 1993, 76–77, quoted in Sheila J. Nayar, "The Values of Fantasy: Indian Popular Cinema through Western Scripts," *Journal of Popular Culture* 31, no. 1 (1997): 74.

30. Smith, *Hollywood Meme*, 124–25.

31. Ibid., 125.

32. Lucia Krämer, "Adaptation in Bollywood," in *The Oxford Handbook of Adaptation Studies*, ed. Thomas Leitch (New York: Oxford University Press, 2017), 261.

33. Smith, *Hollywood Meme*, 14.

34. Patrick Cattrysse, *Descriptive Adaptation Studies: Epistemological and Methodological Issues* (Antwerp: Garant, 2014), 123.

35. Naghmeh Rezaie, "Iranian Adaptations: Overview," *Cross Cultural Adaptations*, http://www.crossculturaladaptations.com/iranian-adaptations/overview/.

2

———

# The State of Contemporary Adaptation

## A Revealing Corpus

Betty Kaklamanidou

*I want to express my sincerest thanks to the following students for voluntarily and without extra credit assisting in the creation of this study's corpus: Alkisti Aktso-glou, Giorgos Dimoglou, Matina Galatsianou, Magda Kolokytha, Konstantinos Ntinias, Sotiria Papantoniou, and Maria Poutiou.*

IN THE "INAUGURAL" BOOK on adaptation studies, George Bluestone cites four different surveys that provide data regarding the use of literature as a source for screenplays. The first examined the output from RKO, Paramount, and Universal between 1934 and 1945 and showed that approximately one-third of feature films were based on novels. The second study inspected "5,807 releases by major studios between 1935 and 1945" to demonstrate that "976 or 17.2 percent were derived from novels." The third revealed that in 1947, "of 463 screenplays in production or awaiting release, slightly less than 40 per cent were adapted from novels." Finally, the fourth was in the form of a *New York Times* article that noted that the "original screenplay" reached "a new low" and "represented only 51.8 per cent of the source material of the 305 pictures reviewed by the Production Code office in 1955."[1] Likewise, Thomas Leitch cites John Tibbetts's more contemporary survey into the 1894–1912 Kemp R. Niver's catalog *Motion Pictures from the Library of Congress Paper Print Collection*. Tibbetts concludes that "of the thousands of titles listed and described, over one-third contain references that demonstrate that they were either derived from stage plays or that they at least in some way simulated the illusion of a theatrical presentation."[2]

The numbers not only are enlightening but represent the kind of primary research that humanities and adaptation studies in particular can greatly benefit from, as data cannot but present the objective situation at a given historical context. Unfortunately, very rare endeavors are mentioned in the relevant literature after the 1960s while most authors tend to repeat that one-third of all Hollywood films come from a literary source, based perhaps on one of the studies presented by Bluestone and Leitch above. Why is this the case? Primary research is important, and especially today, in the era of Web 2.0 and its overwhelming abundance of information, scholars could compile any sort of data suitable to their study. In addition, as digital humanities—another buzzword circulating in the current academic landscape—has been on the rise, the use of technology and its various applications can indeed illuminate areas that were left in the dark or needed a lot of effort to be revealed. Yet the majority of contemporary adaptation monographs and collections—as well as the majority of film studies' publishing output, for that matter—focus on paradigmatic case studies and/or groups of films (the latter usually based on the works by a single author) or methods of analysis and theories and lack information on a larger scale. As the reasons behind the scarcity of specific data in the humanities are varied and extend far beyond the scope of this chapter, the main hindrance is the continuing devaluing of the humanities in the neoliberal context of the last two decades. This results in rare funding opportunities for scholars to undertake this kind of primary research, one that requires long hours and research groups. On the other hand, despite the institutional obstacles, even smaller-scale data compilation can still cast light on persistent questions and should actually be the first step in charting a specific humanities field.

For instance, in 2015, Stephen Hollows conducted research on "the 100 highest grossing films at the US box office in each year between 2005 and 2014 (1,000 films in total)"[3] and concluded that

a.  only 39 percent of the top 100 movies examined were based on original screenplays;

b.  the budget of original films is lower than that of adapted screenplays, remakes, or sequels; and

c.  the main common source remains literature (novels or short stories).

As Hollows did not provide the data online, I wanted not only to corroborate his findings but to undertake new research, adding more films and more explicit criteria. As a trained structuralist, I opted for a quantitative semiotic analysis, a method that "moves from an observed diversity, unifies it from the perspective of constituting the corpus, and finally proves its objectivity via instrumented investigation."[4] In other words, my team and I amassed hundreds of films that we initially grouped under the general label of "film adaptations" before deciding what should be included in the corpus based on selected criteria and proceeded with their statistical analysis and more thorough investigation. That being said, this chapter examines and analyzes a corpus of 802 film adaptations from a total of 1,500 titles released from 2006 to 2015, with a view to providing a comprehensive chart of the current adaptation landscape and answering questions like these: Do most adapted novels belong to the literary canon, or are they mostly popular best sellers? What is the situation with the rise of comic book adaptations in the twenty-first century? Should we consider the numerous sequels and remakes found in the corpus adaptations? And if we do, are they adaptations of the original source (i.e., a comic book or a novel series), or are they basically adapting the first film of their respective franchise? Or both?

The research was undertaken by a group of undergraduate students who happily volunteered to assist me during the summer of 2016 after completing my film adaptation course the previous semester. Each student had to look at the top 150 titles of the US box office provided by boxofficemojo.com and create an Excel file with the following data about each film: title, year of release, director(s), screenwriter(s), source type (in the case of an adaptation), source writer, source year of release/publication, distributor, and worldwide grosses. Once the ten Excel files were collected, I had to confirm the findings, add missing titles, and/or correct minor mistakes and compile the final corpus of film adaptations into a single file.

As with Hollows, this research was not confined to literary novels but included every possible source—from news articles to memoirs and from comics and graphic novels to theater plays and biographies; the corpus includes every film whose script is based on another source, irrespective of type. This seemingly uncomplicated approach, however, raised several difficult questions; for example, were films based on true stories whose credits did not mention a specific nonfiction book/news article/documentary/historical archive considered adaptations or not? For instance, based on its credits, all major online databases, and award institutions, *Selma* (2014), the acclaimed historical/political drama about the 1960s voting rights marches in the United States, is considered a film that brings to life an original screenplay by the British writer Paul Webb. Yet, one can be almost certain that the overwhelming majority of its viewers would consider the depiction of events and characters, such as Dr. Martin Luther King Jr. and President Lyndon Johnson, as drawn from the actual historical events of the represented period. In fact, Webb himself has openly noted that when he decided to write the script, he "started reading about the era and about Johnson" as well as "everything [he] could get [his] hands on about Martin Luther King."[5] Thus, Webb based his dialogue on both actual and fictional facts while director Ava DuVernay and her crew created an authentic representation of the era. Questioning why a script like *Selma*'s is considered "original" and not an adaptation of historical sources becomes more complicated when there are contemporary history films, such as *Lincoln* (2012) and *Unbroken* (2014), whose credits state their screenplays are adaptations of nonfiction works. Does the difference between the scripts of *Selma* and *Lincoln* or *Unbroken* lie in budgetary reasons? Were *Selma*'s producers unable to secure the rights to a specific nonfiction/historical work, or did the author refuse to sell the rights or have demands (monetary or other) that could not be met? Did *Selma*'s screenwriter do such exhaustive research that no one book could serve as his main source? Should we therefore label a film as an adaptation despite the absence of explicit mention in the credits, when we are aware of the filmmaking and production teams' documented intention and their study of primary sources, as well as their final execution (the film itself), at least in the case of a history film, which by definition adapts events that have already taken place and presents mostly well-known characters?

Having adhered to Dennis Cutchins's definition of adaptation as "a way of looking at texts,"[6] I decided to create an inclusive rather than an exclusive corpus.

*Selma* can be analyzed, among other ways, as a political film, a major motion picture directed by a woman, a film about the African American struggle for basic civil and human rights, or a history film that adapts a certain momentous event. Interestingly, most media attention focused on this last category, asking how accurate *Selma* was as a history film—despite the film's original screenplay.[7] At the same time, *Lincoln*, a de facto adaptation as its credits acknowledge, ignited similar media debates.[8] Although this discussion merits the word limit of a whole essay or indeed a monograph, it is not part of this chapter, which aims primarily to outline the contemporary field of adaptation sources, present the data and an initial categorization, and see whether any special patterns or trends emerge. At the same time, I try to confirm or refute previously held ideas about adaptation and use the corpus as the basis for posing new questions and answering lingering ones.

## The Corpus: Compilation and First Observations

As mentioned, a total of 1,500 films produced and/or coproduced in the United States were examined to ascertain whether or not they are adaptations. Based on inclusivity, the final corpus contains 802 films whose source comes from another "text" irrespective of medium. In itself, the number of adaptations is impressive: it constitutes 53.4 percent, more than half of the totality of examined films. One would argue that this result is not entirely valid as it privileges commercially successful titles over the whole of distributed films in the United States during the 2006–15 period, which includes between 600 and a little more than 700 titles annually. In addition, the perceived threat and/or dislike of corpora by scholars often provokes such concerns "as the exclusion of marginal examples and canonisation of favourites."[9] Thus, a venture was undertaken to examine all the titles for two given years chosen at random, 2012 and 2015. Having excluded first all foreign productions and documentaries, quite a substantial selection of the films found between the 151st box office spot and the last (i.e., around 90 titles were excluded from numbers 151 to 300 for the year 2015, which amounts to 60 percent), the data revealed that the percentage of adaptations does drop to around 40 percent, decreasing the total number to around 46 percent. However, the 40 percent, based on a very small-scale research that cannot be considered as "binding" as the findings of a ten-year research project, can be used only as a rough estimate suggesting that the number of all distributed adaptations in a year is somewhat less than 50 percent, a figure that can be attributed to budgetary reasons (i.e., low-budgeted films cannot afford rights), yielding a percentage of total adaptation that remains close to half of the film productions in a given year.

The first observation once the final corpus was complete is that the classification of adaptations is a daunting if not impossible task. For instance, *The Danish Girl* (2015) is based on the same-titled 2000 bestseller by David Ebershoff, which is in itself inspired by the real life of Idris Elbe, a Danish transgendered woman. Should we take the film's credit into account and thus treat the adaptation as one originating from a novel, or should we treat both novel and film as historical documents of a life that did not find its way into history books? In which "source column" do we place films whose stories come from the Bible? True stories or fiction? Where do we put a film like *Machete* (2010), which is based on a fake trailer that played in the double feature *Grindhouse* (2007)? Is *Speed Racer* (2008) an adaptation of the 1966 single manga book, its two televised adaptations (in the 1960s and the 2000s), the release of specific chapters of the original manga in the 1990s, or the release of the manga by DC Comics in 2008? What happens with franchises? Should we simply proceed on a single-film-as-adaptation basis, or are all the films of a given series to be taken as adaptations of the same published source (most often a comic book), or is the second film an adaptation of the characters as they were constructed and specifically visualized in the initial film?

The following study presents an effort to group adaptations into eight categories despite the inherent difficulties of the endeavor, using the type of source each film adapts as the main criterion. These groupings are not absolute, and some of them overlap with others. Yet this initial classification is helpful in that each of the eight clusters discussed below invites sometimes unique and other times overlapping questions that merit future investigation in adaptation studies. Finally, the obvious word limitations prevent each category from extending to a chapter in itself. Thus, I limit myself to some introductory observations and a variety of questions that are pertinent or indicative of each category (they range from structuralist, definitional, and analytical to evaluative queries and are by no means exclusively linked to each adaptation group I discuss), hopefully creating room for future discussions, debates, and research.

## The Novel Category

From the 802 titles of the corpus, 283 films are based on novels, novellas, and short stories (171 novels, 6 novellas, and 14 short stories). This number constitutes 35.2 percent of all adaptations and indeed corroborates the constant adage that one-third of all films have literary sources. This high percentage also confirms that there has been a steady literary output—from the first relevant studies in the 1930s and 1940s that Bluestone cited—for almost nine decades that lends plots and characters to the film industry. The 283 sources are further divided between those novels, novellas, and short stories that target adult audiences and those that belong to children's and young adult literature because the latter's number accounted for 39.8 percent (92 children's and young adult novels were found among the totality of the corpus's literary prose). This number confirms Hollywood's continuing investment in the child/teen demographic despite the Motion Picture Association of America's 2017 statistics that show the top three age groups of frequent moviegoers in 2017 were the 25–39 year age group (11.1 million), the 40–49 year age group, and the 60-plus year age group (6.2 million). The 18–24 year age group is in the penultimate place with 5.2 million, coinciding with the 12–17 year age group (5.2 million) and surpassing only the 3–11 year age group (3.3. million).[10] Yet, taken together, the latter three groups, which represent children, teenagers, and young adults, remain Hollywood's favored demographic target.

More than 70 percent of the adapted novels have been published from the 1980s onward; 57 percent are based on novels published after 2000. Likewise for children's and young adult literature: 54 of the 92 sources (58.6 percent) are novels and fairy tales from the 1990s and early 2000s. It is clear that literature is far from "dying," as many would have us believe,[11] but is instead steadily producing works with intricate plots and characters that attract Hollywood's attention. Additionally, book series, especially from the children's and young adult literature section, have produced highly popular film franchises like the *How to Train Your Dragon* films (2010, 2014, 2019), the *Percy Jackson* films (2010, 2013), the *Hunger Games* films (2012–15), the *Harry Potter* films (2001, 2002, 2004, 2005, 2007, 2009, 2010, 2011), and the *Twilight* film series (2010–15), while the "adult" audience has at the same time been enjoying the *Bourne* franchise, the *Fifty Shades* trilogy, and the Bond films.[12]

These last observations, which include film sequels, reflect the ways the novel category overlaps with the sequel and remake category that is examined later on. While I do not consider the overlap an obstacle, its existence raises several interesting questions beginning with the question of what exactly the sequel adapts. The answer is certainly not just a written text. The first film's success dictates that the sequel should follow, apart from the written source's main plot, similarities in visual style, editing rhythm, music composition, photography patterns, actors, and so forth—in other words, it should incorporate and extend elements of the first film. It has become common, for instance, to refer to how James Whale's *Frankenstein* (1931) is as much a source for the many film and TV incarnations of Mary Shelley's novel as the original written text itself.[13] After all, it is Whale's mise-en-scène and the film's particular iconography (the monster's face and attire, the doctor's lab, etc.) that have become synonymous with the myth. Although a number of sequels ignore this rule (i.e., *The Bride of Frankenstein*, 1935), visuality and iconography matter, and examples abound in our time as well with a contemporary twist. Let's take the *Fifty Shades* film trilogy, which is based on E. L. James's controversial novels. The first film, *Fifty Shades of Grey* (2015), introduced us to the leading couple, Christian Grey (Jamie Dornan) and Anastasia Steele (Dakota Johnson), and their sexual adventures. As a powerful and influential entrepreneur, Grey lives in a Seattle apartment that is an impressive paradigm of sleek architectural design and superiority. The same apartment features in all three films, even after Christian and Anastasia get married and buy a new house. In the era of Web 2.0, this setting did not go unnoticed: Grey's apartment even has its own website. The user can go on a virtual tour accompanied by music from the films and relive some of the stories' moments. However trivial this may seem, it does make one wonder how the novel presents the apartment, whether the production designer's choice was influenced by the written word, or how the viewers' reaction and/or admiration of this luxurious space (on social media) led to the eventual use of a film's setting as a marketing tool.

## The Wider Literature Category

By definition, literature includes not only novels and stories but also plays and poetry. These latter materials provide contemporary cinema with a rather limited degree of inspiration. Out of the 802 films, there were thirty plays, including twelve musicals, and two poems (*Beowulf* [2007] is an adaptation of the Old English epic poem, and *The Sorcerer's Apprentice* [2010] draws its inspiration from, among other sources, Johann Wolfgang von Goethe's 1797 poem). This category represents only 3.9 percent of the total adaptation production. An obvious question is, Why has the percentage of movies based on plays declined so dramatically since the 1930s, when the adapted plays, according to Peter Lev's research,[14] numbered more than 110? Is this a random discrepancy, or did specific reasons lead to a decrease of adapted theater? The second observation is that these adaptations present the same dilemma concerning the sources as the novel category. Are some of the titles based on a single source or several? For instance, John Logan based his screenplay of *Sweeny Todd: The Demon Barber of Fleet Street* (2007) on the eponymous 1979 Broadway musical whose book was written by Hugh Wheeler and whose songs and lyrics were written by Steven Sondheim. Yet the musical's book and Wheeler's work were based on Christopher Bond's 1973 play, which in its turn was inspired on the character of Sweeney Todd, who first appeared as a scoundrel in the Victorian popular serial *The String of Pearls*. In addition, Todd appeared in at least four films from the silent era to the 1970s before becoming popular again in the 2000s with Tim Burton's adaptation and Johnny Depp's able performance. What this reference on Todd's lineage shows is that the 2007 film is a hypertext with more than one hypotext,[15] despite the credits claiming that it was based on only one. As there is no information on the screenwriter's process and his knowledge of Sweeney's long life as a literary, stage, and film character before his musical incarnation, we can only presume that some aspects may have found their way into the construction of the 2007 version.

A similar problem is raised with *Nine*, the adaptation of the same-titled 1982 musical (book by Arthur Kopit, music and lyrics by Maury Yeston), which in its turn adapts Federico Fellini's *8½* (1963), whose autobiographical impulses have been well documented in academia.[16] What is an adaptation scholar to do when confronted with cases such as the last two? Could we talk about a double-source, triple-source, or even quadruple-source adaptation? Leitch has already discussed this in 2007 and has shown that "adaptations may be framed by dozens of contexts exerting different kinds of force, but their need to follow the dictates of these texts is qualitatively different from their decision to follow what are often called their source-texts."[17] For Leitch, it is "the source-text" that "provides the primary inspiration for a given adaptation, even though that inspiration may be inflected or deflected by any number of variously exigent contexts."[18] Thus, in the case of *Nine*, one should primarily focus on the film's relationship with the musical while taking the other texts that inform the adaptation (both intra- and extratextual) as contextualizing factors. On the other hand, as Leitch astutely observes, the

> attempts to impose order on the endless web of intertextuality institutionalizing some intertexts as texts and marginalizing others as contexts is itself institutional, ad hoc, and subject to ceaseless revision. What makes a context into a text is the act of being treated as a text by somebody; what demotes a text to a context . . . is being rejected as a text.[19]

Agreeing with Leitch, I would add that although the search for the "ultimate" original text can be a futile endeavor, it can bring to light obscure and forgotten texts and authors. At the same time, one should not forget that the textual sources Western society has been celebrating for years, decades, or even centuries are always determined—beyond inherent values and characteristics—by sociohistorical circumstances and that their popularity remains as long their context allows for their continued acclaim.

## The Versions of Truth Category

More than 16.5 percent of all adaptations are based on real people and events. For this category, I coined the label "versions of truth," because I agree with Roland Barthes that "historical discourse is a fake performative discourse in which the apparent constative (descriptive) is in fact only the signifier of the speech-act as an act of authority."[20] Following this thesis, the 133 films of the corpus that adapt sources from this category—memoirs, biographies, and autobiographies; historical documents; news articles; and documentaries—do not exactly present the "truth" but rather versions of it, while retaining what Barthes call a "reality effect."[21]

In a further subdivision, this category includes two documentaries that served as the basis for two fiction films and 131 nonfiction books. The acclaimed documentary *Planet B-Body* (2007) served as the basis for the dance film *Battle of the Year* (2013), and *The Fighter* (2010) is the dramatic adaptation of the 1995 documentary *High on Crack Street: Lost Lives in Lowell*. The nonfiction books can be further subcategorized into biographies, autobiographies, historical books, personal memoirs, and even self-help manuals and business books. The rom-com *He's Just Not That Into You* (2009) is based on the same-titled self-help book by Greg Behrendt (2004), while the sports drama *Moneyball* (2011) is the adaptation of the Michael Lewis business book *Moneyball: The Art of Winning an Unfair Game* (2003). Again, this category overlaps with the novel category. *The Other Boleyn Girl* (2010) is an adaptation of Philippa's Gregory historical novel, which in turn is based on historical facts about Henry VIII's second wife.

Another observation regarding the versions of truth category is that its films belong to a variety of genres—from political satires (*The Men Who Stare at Goats*, 2009) and comedies (*Pitch Perfect*, 2012) to biopics (*The Wolf of Wall Street*, 2013), history films (*The Conspirator*, 2011), and rom-coms (*Think Like a Man*, 2012)—since their only link is that they dramatize a type of truth (historical, personal, or otherwise). Most films emphasize their correlation to reality with the tag "based on a true story" that accompanies their opening credits. As Leitch observes, this label "seems to have come into common use only during the 1990s" with films such as "*Awakenings* (1990) and *GoodFellas* (1990)," although he adds that "the concept is much older" and notes that "*Dog Day Afternoon* (1975) begins with the title, 'What you are about to see is true—it happened in Brooklyn, New York, on August 22, 1972.'"[22]

Complications arise between those films that explicitly state they are adaptations of previously published material (*Charlie Wilson's War*, 2007; *12 Years a Slave*, 2013) and those that do not include their sources in their credits and whose screenplays are considered "original" (*The Conspirator*, *Frost/Nixon*). Is *12 Years a Slave*, based on Nathaniel Northop's memoir, a "better" film than *The Conspirator*, whose original screenplay took its writer, James D. Solomon, eighteen years of research, writing, and rewriting[23] and presents a historical event largely unknown to most American and global viewers: the expedited trial and subsequent hanging of Mary Surratt on charges of conspiracy to assassinate President Lincoln. Similarly, although writer Dustin Lance Black has acknowledged that he conducted extensive research for the screenplay of *J. Edgar* (2011),[24] no source is mentioned in the film's credits.

Leitch argues that "the phrase 'based on a true story' . . . appeals to the authority of a master text that has all the authority of a precursor novel or play or story with none of their drawbacks. . . . [I]ts authority can never be discredited."[25] On the other hand, Genevieve Koski, Keith Phipps, and Tasha Robinson propose examining the label "based on a true story" as "a marketing term," since "every narrative film is a work of fiction to some degree."[26] Be it a marketing tool or a reliable master authority (or both), the "truth" as an avowed source seems to be gaining traction in the field of adaptation. As reality television has managed to become the most watched genre on television, from *Survivor* (CBS 2000–present) to *American Idol* (Fox 2002–16, ABC 2018–present) and *Keeping Up with the Kardashians* (E! 2007–present), are we witnessing a similar trend in cinema?[27]

## The Sequel and Remake Category

Although sequels and remakes can be taken as two distinct categories, I group them together for three reasons: first, the source material for both is a previously distributed audiovisual text, thus constituting a different type of adaptation; second, both groups have been invariably deemed in the medium's history as parasitic,[28] marking the death of originality and the proof of Hollywood's lack of imagination and/or creative laziness;[29] and third, both sequels and remakes (along with spin-offs, reboots, and prequels) "are best understood as historical varieties of a serial practice that is distinct to Hollywood's commercial film culture."[30]

From the 802 corpus films, 63 are remakes (7.8 percent). Leitch remarks that "remakes differ from other adaptations to a new medium and translations to a new language because of the triangular relationship they establish among themselves, the original film they remake, and the property on which both films are based."[31] Many of the 63 remakes are based on screenplays, which in their turn have adapted a novel (e.g., both the British *The Wicker Man* [1977] and its same-titled 2006 remake are inspired by David Pinner's 1967 novel *The Ritual*); a play (e.g., *The Women* [1939] and its remake *The Women* [2008] are based on Clare Boothe Luce's 1936 play); or a short story (e.g., *Total Recall* and its 2012 remake both adapt Phillip K. Dick's 1966 short story *We Can Remember It for You Wholesale*). In this case, as Leitch adds, "the producers of a remake typically pay no adaptation fees to the makers of the original film, but rather purchase adaptation rights from the authors of the property on which that film was based, even though the remake is competing much more directly with the original film than with the story or play or novel on which both of them are based."[32] Yet our remake list also contains several titles that are only based on a previous film. *The Hills Have Eyes* (2006) is a remake of the 1977 original screenplay; *Delivery Man* (2013) is a remake of the 2011 French Canadian *Starbuck*; and *Fright Night* (2011) is based on the screenplay for the same-titled 1985 horror film. However, the process of acquiring the rights to the original film is as lengthy as in the case of a first adapted screenplay. According to entertainment lawyer Robert L. Seigel,[33] the process starts with checking "the copyright notice on the original film," which "may have been transferred one or more times over the years." Then,

> you would have to check with the current owner of the rights in and to the original film to determine if it or someone else has the right to remake the original film since some distributors only have the rights to the original film itself especially if the distributor acquired the rights to the film from a third party and it did not finance and produce the film itself. Furthermore, you would have to determine who owns the rights to the screenplay of the original film.[34]

When viewed as adaptations, remakes can potentially be very useful in displaying the change in the visual aspect of the text, something that is frequently neglected in adaptation studies. Scholars can profitably examine technological advances, new camera and editing trends, and their function in the text. How did the remake imagine the fictional cosmos? Since all remakes provide different experiences from the films they originate from, if only because they frame themselves as remakes, how important is it to establish whether they seek primarily to emphasize or to efface these differences? What was the reaction of the young audience to the remake? After all, despite the potential use of the word "remake" or the allusion to the original in the film's promotional material for the new viewers, the remake may well be their "original": for most GenXers in Greece, James Bond will always be Roger Moore. Finally, what was the reaction of the viewers who are old enough to have watched the first film and are now privileged to be present for its next edition? As Verevis rightly notes, remakes are contingent on an active audience that has "not only prior knowledge of previous texts and intertextual relationships, but an understanding of broader generic structures and categories."[35]

Similar complications arise with sequels. This adaptation type outnumbers remakes, accounting for 19.5 percent of all 802 adaptations in our corpus, with 157 titles produced and released between 2005 and 2016. Although remakes and sequels can be discussed together, there are differences among them. Carolyn Jess-Cooke and Constantine Verevis note that "the sequel does not prioritize the repetition of an original, but rather advances an exploration of alternatives, differences, and reenactments that are discretely charged with the various ways in which we may reread, remember, or return to a source."[36] Apart from being one of the most significant parts of a crossover, multiplatform universe with possibilities for huge revenues, sequels follow the postmodern tendency toward serial repetition, examined by Umberto Eco in the 1980s.[37] Despite the implications of contemporary serialization in both twenty-first-century film and television, sequels, spin-offs, reboots, and prequels are important in adaptation studies insofar as we can examine them as textual opportunities to question the limits of adaptation practices.

## Comics, Graphic Novels, and Mangas

This grouping includes 36 films (4.4 percent) that borrow storylines and characters from comic strips, comic books, graphic novels, and mangas. Although comics had been regularly adapted since the 1930s, the explosion of similar adaptations since the 2000s on both film and television as well as the evolution of adaptation studies have rendered this type of adaptation a distinct subject of interest.[38] This category overlaps to an extent with the sequel and remake category, as it contains some titles that constitute the first films of a given franchise (*Iron Man*, 2008; *Thor*, 2011; *Captain America: The First Avenger*, 2011) and films that were originally envisioned as the first part of a film series that proceeded no further (*Whiteout*, 2009; *The Losers*, 2010).

Yet, taken together, comic adaptations along with their sequels and remakes led to the characterization of the previous decade as "The Golden Age of Comic Book Filmmaking,"[39] and one could safely hypothesize that the second decade of the 2000s will also bear the same label. The sheer number, global reach, and impressive revenue of comic adaptations invite a number of questions, such as the reasons behind the phenomenon, mainly the cultural, sociopolitical, and industrial contexts that keep on investing in these narratives. For adaptation studies, the transformation of the pictorial character of comics (panel drawings with speech bubbles) is only one possible focus of the yet limited literature, which has generally avoided questions about what happens to visuality and iconography during the adaptation process or how cinema has been influenced by comics' visual character.[40] Questions like these can lead to new paths not only for the field but for film studies as well.

## TV Series

Thirty-six films between 2006 and 2015 were inspired by television (4.5 percent of all 802 adaptations). Most films adapted a TV series, but there were several adaptations of specific TV episodes (*Real Steel*, 2011) and TV movies (*Don't Be Afraid of the Dark*, 2010). The majority of the television narratives that found their way to the big screen are action/adventure and/or sci-fi stories, such as the film versions of *Miami Vice* (NBC, 1984–90), *Mission: Impossible* (CBS, 1966–73), the *Star Trek* six television series, and *21 Jump Street* (Fox, 1987–91), confirming Hollywood's blockbuster policy.

Although popular television series have an established following, this does not by definition guarantee the successful financial course of their adaptation. Despite the impressive box office results of the *Mission: Impossible* franchise (the five films of the series that began in the 1990s have had gross profits of almost $2.8 billion worldwide, with each film averaging around $600 million), the *Star Trek* films (with grosses of around $2 billion worldwide, not counting the sizeable profits from relevant merchandise), as well as the two *Sex and the City* installments, which amassed more than $703 million on a combined budget of $165 million, not all beloved series become commercially successful. For instance, *Miami Vice* (2006) cost $135 million but grossed only $163 million worldwide. Despite the great popularity of the NBC same-titled show (1984–90) and the positive reviews, the film did not manage to attract viewers to theaters. Similarly, the adaptation of NBC's *The A-Team* (1983–87), a film that had been in development for many years and was finally released in 2010, was budgeted at \$110 million but managed to bring in only \$177 million despite Liam Neeson's and Bradley Cooper's star power. What were the mistakes of these latter productions?

Sandy Schaefer[41] notes that marketing strategies and release dates must be taken into account when examining a given film's success or failure at the box office. To elucidate her thesis, Schaefer discusses the publicity strategies behind two highly anticipated 2015 titles, *Tomorrowland* and *Mad Max: Fury Road*. She argues that the first's disappointing number of admissions and the second's unprecedented monetary achievement are rooted in the decision not to market *Tomorrowland* as a blockbuster linked to Disney's theme park but instead to focus on George Clooney's brand and in the strategy to market *Mad Max: Fury Road* in a way that created positive word-of-mouth prior to its release.

## Video Games

The results from the corpus show that video game adaptations constitute only 1.2 percent of the total adaptations in the 150 films of the worldwide box office from 2006 to 2015. The 10 films based on video games are *Silent Hill* (2006), *Resident Evil: Extinction* (2007), *Hitman* (2007), *Prince of Persia: The Sands of Time* (2010), *Resident Evil: Afterlife* (2010), *Resident Evil: Retribution* (2012), *Silent Hill: Revelation* (2012), *Need for Speed* (2014), *Ouija* (2014), and *Hitman: Agent 47* (2015). These films brought their respective studios more than \$1.6 billion in revenue with a total cost of about \$575 million, a profit margin that underlines, on the one hand, an already established audience—mainly gamers who will watch the cinematic adaptation irrespective of its stars, director, story, and marketing—and, on the other, the not (yet?) founded popularity of this adaptation.

The majority of video game adaptations have been harshly criticized in the media, while their box office success varies. Interestingly, star bankability and video game popularity do not always produce economic rewards. Jake Gyllenhaal starred in the adaptation of the video game franchise *Prince of Persia*, directed by Mike Newell, a seasoned director with dozens of credits, including *Four Weddings and a Funeral* (1994), *Mona Lisa Smile* (2003), and *Harry Potter and the Goblet of Fire* (2005). Yet the high-budget adaptation, which cost \$200 million, returned only \$336 million worldwide, prompting Brandon Gray to rank the film among the big "botches" of 2010.[42] This example stresses the importance of further study into this new form of adaptation but also invites the examination of how the video game culture has inspired films to be adapted in video games. Based on a rigorous study, Alexis Blanchet shows that the last four decades witnessed an increase in the production of video games that are adapted from films. Specifically, Blanchet notes that "from 1975 to 2010, 547 films shown in movie theatres gave rise to around 2,000 games," adding that this practice "over the last 35 years" has "become a major category in video game production today, accounting for nearly 10 percent of the total number of video games published."[43]

## Minor Categories and "Impossible" Adaptations

The corpus's adaptations include, among others, nine adaptations based on toys—*Transformers* (2007), *Transformers: Revenge of the Fallen* (2009), *Transformers: Dark of the Moon* (2011), *Transformers: Age of Extinction* (2014), *G.I. Joe: The Rise of Cobra* (2009), *G.I. Joe: Retaliation* (2013), *The Lego Movie* (2014), *Ouija* (2014), and *Bratz* (2007); seven parodies that adapt a number of film texts (e.g., *Date Movie* and *Scary Movie 4*, both 2006); and the three Alvin and the Chipmunks films based on the animated music group.

*Oblivion* (2013) is a case of what I call an "impossible" adaptation. According to the film's IMDb page, the screenplay is based on Joseph Kosinski's graphic novel. Yet, Rich Johnston[44] states online that the comic book was never published and all that exists are a few concept pages. Johnston concludes that the media focus on the comic origins of the film was probably just a marketing ploy to attract comic aficionados to the theater.

To conclude, I'd like to point out the following. First, the preceding statistical analysis confirms three statements that are widely reiterated in adaptation studies: first, more than half of Hollywood's film productions are still heavily dependent on adaptations; second, one-third of these adaptations are based on novels; third, new and/or relatively new types of adaptations (e.g., comics and television) can have steady and sizeable returns. Yet, the data collection undertaken here and its initial analysis offer something more important: a certain mapping of the adaptation field at a specific historical juncture that not only reveals its width and importance to film studies but also clearly delineates its major subsections, thus providing both a general outline and different objects of analysis. Both isolated case study examinations as well as wider historical and interpretative studies on adaptation can greatly benefit from the statistics presented here, even if one chooses to place their chosen subject in their place of Hollywood's adaptation preferences only in the decade examined.

## Notes

1. George Bluestone, *Films into Novels* (Baltimore: Johns Hopkins University Press, 1957), 3.

2. Thomas Leitch, *Film Adaptation and Its Discontents* (Baltimore: Johns Hopkins University Press, 2007), 23.

3. Stephen Hollows, "How Original Are Hollywood Movies?" *Stephenhollows.com*, June 8, 2015, https://stephenfollows.com/how-original-are-hollywood-movies/.

4. François Rastier, "Computer-Assisted Interpretation of Semiotic Corpora," in *Quantitative Semiotic Analysis*, ed. Dario Compagno (Cham, Switzerland: Springer, 2018), 129.

5. Quoted in Vincent Dowd, "Oscars: Selma Writer Tells His Side of Row with Director," *BBC*, February 20, 2015, https://www.bbc.com/news/entertainment-arts -31539526.

6. Dennis Cutchins, "Bakhtin, Intertextuality, and Adaptation," in *The Oxford Handbook of Adaptation Studies*, ed. Thomas Leitch (Oxford: Oxford University Press, 2017), 79.

7. See Dee Lockett, "How Accurate Is *Selma*?" *Slate*, December 14, 2014, http://www.slate.com/blogs/browbeat/2014/12/24/selma_fact_vs_fiction_how_true _ava_duvernay_s_new_movie_is_to_the_1965_marches.html?via=gdpr-consent; Caitlin Gibson, "How Accurate Is 'Selma'?" *Washington Post*, February 17, 2015, https://www.washingtonpost.com/news/arts-and-entertainment/wp/2015/02/17/ how-accurate-is-selma/?utm_term=.bbf14bbdc702 (accessed March 20, 2017); and Alex von Tunzelmann, "Is Selma Historically Accurate?" *The Guardian*, February 12, 2015, https://www.theguardian.com/film/2015/feb/12/reel-history-selma -film-historically-accurate-martin-luther-king-lyndon-johnson.

8. See Forrest Wickman, "How Accurate Is Lincoln?" *Slate*, November 9, 2012, http://www.slate.com/blogs/browbeat/2012/11/09/lincoln_historical_accuracy_sorting _fact_from_fiction_in_the_steven_spielberg.html; Hendrik Hertzberg, "'Lincoln'

v. Lincoln," *New Yorker*, December 17, 2012, https://www.newyorker.com/news/hendrik-hertzberg/lincoln-v-lincoln; Harold Holzer, "What's True and False in 'Lincoln' Movie," *Daily Beast*, November 22, 2012, https://www.thedailybeast.com/author/harold-holzer; and von Tunzelmann, "Is Selma Historically Accurate?"

9. Constantine Verevis, *Film Remakes* (Edinburgh: Edinburgh University Press, 2005), 2.

10. Motion Picture Association of America (MPPA), "Theme Report, 2017," https://pmcdeadline2.files.wordpress.com/2018/04/mpaa-theme-report-2017.pdf.

11. See, for instance, Will Self, "The Novel Is Dead (This Time It's for Real)," *The Guardian*, July 18, 2013, https://www.theguardian.com/books/2014/may/02/will-self-novel-dead-literary-fiction; and Joel Breuklander, "Straight, White Guys, At Least," *The Atlantic*, 18 July 2013, https://www.theatlantic.com/entertainment/archive/2013/07/literature-is-dead-according-to-straight-white-guys-at-least/277906/.

12. The division between novels for adults and novels for young adults is made based on how the titles are marketed and does not mean either categories cannot be read and/or enjoyed by readers of both age groups.

13. See Dennis R. Perry, "The Recombinant Mystery of Frankenstein: Experiments in Film Adaptation," in Leitch, *Oxford Handbook of Adaptation Studies*, 137–53.

14. Peter Lev, "How to Write Adaptation History," in Leitch, *Oxford Handbook of Adaptation Studies*, 661–78.

15. According to Gérard Genette's five categories of intertextual relations, a hypotext can be considered as the adaptation source, "an anterior text A," and the hypertext is the adaptation itself, or "text B," Gérard Genette, *Palimpsestes: La littérature au second degré* (Paris: Seuil, 1982), 11–12.

16. See, for instance, Peter Bondanella, *The Films of Federico Fellini* (Cambridge: Cambridge University Press, 2002); and Bert Cardullo, *European Directors and Their Films: Essays on Cinema* (Lanham, MD: Scarecrow Press, 2012).

17. Thomas Leitch, "The Texts behind *The Killers*," in *Twentieth-Century American Fiction on Screen*, ed. R. Barton Palmer (Cambridge: Cambridge University Press, 2007), 41.

18. Ibid.

19. Ibid., 42–43.

20. Roland Barthes, *The Rustle of Language*, trans. Richard Howard (Berkeley: University of California Press, [1967] 1989), 139.

21. Ibid.

22. Leitch, *Film Adaptation and Its Discontents*, 280.

23. Kiko Martinez, "James D. Solomon: *The Conspirator*," *CineSnob*, April 29, 2011, http://www.cinesnob.net/james-d-solomon-the-conspirator/.

24. Anne Thompson, "Dustin Lance Black Talks Controversial J. Edgar Script: 'We Didn't Put Hoover in a Ball Gown,'" *IndieWire*, November 23, 2011, http://www.indiewire.com/2011/11/dustin-lance-black-talks-controversial-j-edgar-script-we-didnt-put-hoover-in-a-ball-gown-184000/.

25. Leitch, *Film Adaptation and Its Discontents*, 289.

26. Genevieve Koski, Keith Phipps, and Tasha Robinson, "The Problem with 'Based on a True Story,'" *The Dissolve*, September 16, 2013, https://thedissolve.com/features/the-conversation/156-the-problem-with-based-on-a-true-story/.

27. For more, see Davinia Thornley, ed., *True Event Adaptation* (New York: Palgrave, 2018).

28. Leitch, *Film Adaptation and Its Discontents*.

29. See Verevis, *Film Remakes*, 2; and Carolyn Jess-Cooke and Constantine Verevis, *Second Takes: Critical Approaches to the Film Sequel* (Albany: SUNY Press, 2010).

30. Frank Kelleter and Kathleen Loock, "Hollywood Remaking as Second-Order Serialization," in *Media of Serial Narrative*, ed. Frank Kelleter (Columbus: Ohio State University Press, 2017), 126.

31. Thomas Leitch, "Twice-Told Tales: The Rhetoric of the Remake," in *Dead Ringers. The Remake in Theory and Practice*, ed. Jennifer Forrest (Albany: SUNY Press, 2002), 39.

32. Ibid.

33. Robert L. Seigel, "How Do You Secure the Rights for a Remake of an Old Movie?," *MovieOutline*, http://www.movieoutline.com/articles/how-do-you-secure-the-rights-for-a-remake-of-an-old-movie.html.

34. Ibid.

35. Verevis, *Film Remakes*, 2.

36. Carolyn Jess-Cooke and Constantine Verevis, "Introduction," in Jess-Cooke and Verevis, *Second Takes*, 5.

37. In Jess-Cooke and Verevis, "Introduction."

38. See Dru Jeffries, *Comic Book Film Style: Cinema at 24 Panels per Second* (Austin: University of Texas Press, 2017); Liam Burke, *The Comic Book Film Adaptation: Exploring Modern Hollywood's Leading Genre* (Jackson: University of Mississippi

Press, 2016); and Benoit Mitaine et al., *Comics and Adaptation* (Jackson: University of Mississippi Press, 2018).

39. Burke, *Comic Book Film Adaptation*, 23.

40. I should note that recent literature has started to focus on the issue of visuality in adaptation. See Barry Keith Grant and Scott Henderson, eds., *Comics and Pop Culture: Adaptation from Panel to Frame* (Austin: University of Texas Press, 2019).

41. Sandy Schaefer, "Are Original Movies Really a Hard Sell?" *Screenrant.com*, June 26, 2015, http://screenrant.com/tomorrowland-box-office-original-movies-sequels/.

42. Brandon Gray, "'Gulliver,' 'Persia,' 'Narnia' Rank among the Big Botches of 2010," *Boxofficemojo.com*, January 20, 2011, http://www.boxofficemojo.com/news/?id= 3051&p=.htm.

43. Alexis Blanchet, "A Statistical Analysis of the Adaptations of Films into Video Games," *Inaglobal.fr*, December 7, 2011, http://www.inaglobal.fr/en/video-games/ article/statistical-analysis-adaptation-films-video-games (accessed March 20, 2017).

44. Rich Johnston, "*Oblivion*, Based on the Non-Existing Graphic Novel," *bleedingcool. com*, April 12, 2013, https://www.bleedingcool.com/2013/04/12/oblivion-based-on -the-non-existing-graphic-novel/.

# The Impact of Censorship on Adaptations

## The Case of Portugal during the New State (1933–74)

Eurydice Da Silva

IT IS A COMMONLY accepted premise in the adaptation of novels to film that "changes are inevitable the moment one abandons the linguistic for the visual medium."[1] The differences are not only due to the different nature of both literature and cinema but also because each work of art expresses its artist's vision through the specificities of the medium used. Bluestone rightly highlights that the "filmist becomes not a translator for an established author, but a new author in his own light."[2] This observation, however, delves into another thorny topic, that of authorship in film. If the artist behind a novel is the writer, the author of a film is the director and not the screenwriter, at least according to the popular auteur theory, the brainchild of French 1950s criticism.

Jack Boozer suggests that the contemporary perception of cinematic authorship should be taken into account in adaptation analyses, by including extratextual data such as the screenplay.[3] Adaptation studies must encompass a broader reality when it comes to film because from the early steps of development to postproduction, the creative process of the latter is almost invariably subjected to external scrutiny and intrusive input. Unlike the writer's relationship with their novel, the filmmaker's relationship to their film is less direct. This redefines the specificity of adaptation when it comes to reinventing a novel into a film, and a transdisciplinary approach becomes necessary to apprehend the multifaceted reality encompassed by such an adaptation.

French historian Marc Ferro writes that to fully comprehend a film and the reality that it represents, it is necessary to look beyond the images and the narrative beyond the frame.[4] Behind the screen lies everything that is part of the film but that is not the film: the author, the production, the public, and the regime. These extra-filmic parameters have a defining role in shaping the film's DNA, as cinema is an art form that is inherently conditioned by economic factors and the key players involved. It follows that film adaptations are also impacted by these external factors and that the final result does not exclusively rely on choices of artistic nature but is likewise conditioned by the production context.

Nevertheless, the ways that the production context impacts film content remain a rare subject in adaptation studies. For example, although works on censorship abound, the literature on censorship and adaptations is limited and mainly consists of article-length case studies, focused especially on the Hollywood Production Code era.[5]

Rather than overlooking or dismissing nonvisible sources when comparing two different forms of storytelling, we can use them to reveal new meaning and to bridge the gap produced by the process of adaptation from one medium to the other. What could be seen as discrepancies or infidelities between the original work and its adaptation, whether they are of a narrative or aesthetic nature, become a means of decoding the context within which a film is made. Extra-filmic sources linked with the film itself can help establish a dialogue between the two periods of time during which each piece was produced, the original work and its adaptation. The process of adaptation can help us understand a given historical context or further explore our knowledge of it through a different angle, shedding light on methods and practices that are attached to a certain era. If the production environment conditions the content of the film, inversely the film can also inform us on the extra narrative data that led to its confection, thus providing insight into history through cinema. Questions that we can therefore ask ourselves are these: Do the storytelling differences result from the process of adaptation from one medium to the other?

Are they creative choices, or are they the result of imposed external factors such as sociopolitical or economical patterns in a given context?

The latter becomes more evident in an extreme political context such as an authoritarian regime, where interventionist policies from state funding bodies or censorship can orientate film content. This chapter focuses on a particular authoritarian regime, the Portuguese New State (1933–74), and the impact of censorship on film adaptations. It is based on historical research conducted at the French Film Library in Paris, the Portuguese Film Library, and the National Archives of Torre do Tombo in Lisbon, where previously unstudied primary sources regarding Portuguese film history are kept. Regarding film censorship, we adopt here Daniel Biltereyst and Roel Vande Winkel's definition, which views censorship as "the attempt to hinder or limit the free expression, creation, production, distribution, exhibition, and reception of films."[6]

## Portugal: Film Production and the Role of Censorship

From 1926 to 1974, Portugal was ruled by the longest dictatorship in Europe during the twentieth century. On May 28, 1926, a military coup put an end to the First Republic of Portugal. The founder of the next political order, António Oliveira Salazar (1889–1970), gradually came to power, serving as prime minister from 1932 to 1968. On April 11, 1933, he implemented a new constitution and created the New State, a regime that lasted until 1974. From the early days of this authoritarian regime, the assertion of the New State's power was enacted through the establishment of control structures such as the Office of National Propaganda,[7] created on October 26, 1933, which promoted and supervised all matters related to film. All productions, both national and foreign, were submitted to censorship, as were all forms of expression such as press, radio, theater, or literature. The censorship committee wanted to secure the political neutrality of film content and to forbid subversive themes considered to go against the ideology of the New State, based on the founding trilogy of "God, Nation, Family." For over forty years, the legislation relative to film and censorship evolved along with the regime cycles,[8] and film production and history remained strongly entwined. The first censorship decree regarding cinema dates back to February 16, 1925, prior to the military coup, and advocates that "the exhibition of films that go against morality is forbidden."[9] The notion of morality must be comprehended within the context of a country that is deeply rooted in a Catholic tradition. After the 1926 military coup, however, we notice that film censorship gradually gains a new dimension as it also becomes political. An additional decree on May 6, 1927, states that

> the exhibition of the following films is strictly forbidden: films which are pernicious to the education of the people, which incite crime, which are offensive to morality and the political and social regime in place. Films which distinctly present scenes with: mistreatment of women / torture of men and animals / naked characters / lewd dances / surgical operations / capital punishment / brothels / murders / thievery or burglaries with violation of property, where you can see, through the representation of details, the means used to commit such a crime / the glorification of crime through letters or photographic effects.[10]

Writers, directors, and producers found themselves confronted with the problematic interpretation of such a law, which obscurely asserted that anything that was considered "offensive" to the "regime in place" was forbidden. In fact, a wider range of forbidden subjects resulted from the New State's cultural policy and ideology. Although not directly mentioned in the decree, many other subjects that went against the values of the Church and could be viewed as against the general term of "morality," such as suicide, abortion, extramarital relationships, and homosexuality, faced disapproval. Politically, themes subjected to censorship ranged from communism, which was seen as a growing threat and a potential opponent to the New State regime, to criticism against Portugal's national and international policies during the colonial war between 1961 and 1974.

The director of the Secretary of National Propaganda, António Ferro (1895–1956), was instrumental in forging the New State's cultural policy in the early days of the regime. Ferro believed that national cinema was key in forming an ideal vision of Portuguese identity, thus contributing to the portrayal of official stability onscreen. In a speech delivered on December 30, 1947—and later published in 1950—during a film awards ceremony organized by the Secretary of National Information, he shared what he thought ought to be the mission of Portuguese cinema: "Portuguese cinema, in effect, has, among others, two great noble missions: a high educational mission within the Country (in the esthetic and moral sense) and a difficult external mission of bringing to other People the knowledge of our life, our character, and our degree of civilization."[11] In the early days of the regime, Ferro believed that cinema's main vocation should be an educational one. Its role was to inform but also to shape or politically orientate viewers. In 1934, a year after the instauration of the new regime, Ferro published a series of interviews he conducted with Salazar, where he quotes the leader's words regarding the government's mission to secure a path of political stability: "The work to accomplish, especially in this time of national rebirth, must originate in an act of faith regarding the Portuguese Nation and take its inspiration in a healthy nationalism,"[12] a vision that would shape Portugal's cultural policy for the years to come. Cinema was one of the instruments that would help reach that goal. From the beginning of the New State regime, the Office of National Propaganda endeavored to promote and glorify the action of the state mainly through propaganda documentaries with nationalist interests at their core.[13] Apart from documentaries, only two propaganda fiction films were produced in Portugal during the New State regime (1933–74), *The May Revolution* (1937) and *Spell of the Empire* (1940), both directed by António Lopes Ribeiro. As neither of them met commercial success, this must have discouraged the New State from investing further in fiction[14] and led to favoring documentary films as a more viable form of politically educational cinema. The rest of the fiction film production in Portugal between 1933 and 1974 bears the marks of censorship. Whenever the censors applied cuts to screenplays or shot reels, the result was the orientation of narration toward an accepted norm. The cuts that were imposed informed the screenwriter of what was considered acceptable or not, thus suggesting what was deemed appropriate through the unsaid. The result is a body of cinematic work (with a production that ranged from zero to five national feature films released each year) that avoids sensitive themes or ends up addressing them in an indirect manner, a strategy best exemplified in the Portuguese auteur cinema from the 1960s (Cinema Novo or Novo Cinema).

In the first decades of the regime, it seems film adaptations were particularly appreciated by representative figures of the New State. In a speech delivered on August 12, 1946, Ferro as the director of the Office of National Information encouraged writers and directors to favor adapting screenplays based on classics of Portuguese literature, structured stories with proven success, rather than writing original texts. According to him, "Literature (and Americans have understood this) is the greatest source of inspiration for cinema."[15] Adapting celebrated novels allowed screenwriters to avoid "anecdotes without consistence,"[16] an ill that he deemed recurrent in national cinema, in order to follow secure, viable plots.

Concurrently, historical films or biopics based on important figures of Portuguese cultural heritage were also greatly appreciated. For instance, in 1936, *Bocage*, a film about the eighteenth-century Portuguese poet Rocha Martins, was released. A decade later, *Camões* (1946), based on the life of the illustrious Portuguese Renaissance poet Luís de Camões, was chosen by the Office of National Information to represent the country at the Cannes Film Festival. The archives of the French Film Library hold press releases of the time, which reveal how the Portuguese officials wished the film to be perceived abroad. The summary of the film written by the official Portuguese delegation in France describes the main character in the following terms: "Camões incarnates the essential qualities of the Portuguese People: he is brave, audacious, a sincere lover, somewhat frivolous. He sings the 'xacaras,' which are the ancestors of today's fado." The end of the film presentation also echoes this opening, where Camões is instrumentalized in order to embody the patriotic exaltation of the New State:

> In Lisbon, Camoes, solitary, abandoned and ill, is agonizing. Everything that he loved is dead: his friends, his lovers. Even the homeland seems to be dying. . . . In the African sands, the Portuguese flag is trampled by Moorish horses. . . . But the poet has a sudden burst of hope: "No, the Homeland will not die; it cannot die!" From his lips rise up the immortal verses from his poem, which assert the immortality of his Homeland. The poet was right: 1640–1910—the independence is guaranteed. In 1940, Portugal could celebrate its eight centuries of history.[17]

This marketing document reveals the intentions of the country's officials: promote a strong national identity and a stable unity and evolution between past, present, and future and consequently place the New State regime in a natural political continuity, from Portugal's independence from Spain in 1640, to the end of the monarchy and the beginning of the First Republic in 1910, and, finally, to the celebration of Portuguese history in 1940, despite the political coup in 1926 and the instauration of a new order in 1933. The nationalist-oriented dialogue is attributed to a key figure of Portuguese history. This transposition turns Camões into a contemporary spokesperson for the New State's ideals. Ferro defends the quality of this film as well as the nature of its subject:

> Camões marks a milestone in the history of our cinema, which saw *big* for the first time, and dared to tackle, without diminishing it, one of the greatest figures of our History, of our past. A hero and a soldier of another time, one of these real figures like Shakespeare or Cervantes, but who seem legendary with the projection of centuries, yet still true and incredible, because their truth was, above all, like a lived dream.[18]

It becomes clear that to Ferro, cinema can impact reality. Therefore, it should not seek to merely represent life as it is but should instead represent the ideal that the Portuguese people should aim to reach.

The censorship commission made sure that their process remained unnoticed from the public and was mainly characterized by its invisibility. If the cuts on a foreign film were too apparent, the censors would forbid its release altogether. One had to forget censorship even existed to accept it as a norm. As for national films, dialogue and negotiation between censors and directors during preproduction and postproduction would prevent a final version where cuts and changes would be too recognizable.

Cinematic representation of the lives of bygone cultural figures forged a strong image of national identity and therefore of lasting stability. By extension, adapting the works of such important figures was deemed beneficial. This strategy contributed equally to Portugal's glorification as well as to its national and international acclaim. However, if we take a closer look at Portugal's film production, based on Horta e Costa's[19] list of subsidized films produced in Portugal between 1918 to 1948 during the first two decades of the regime, there is no increase in the number of adaptations.

From 1918 to 1932, before the instauration of the New State, fifty-five feature films were produced; twenty-one (38 percent) were adaptations (of novels, plays, or tales). From 1933 to 1948, when António Ferro left the Office of National

Information, the national production of fiction features decreased, with only forty-eight titles produced. Twelve films were adaptations (25 percent of the entire production); four were adaptations of plays, and eight were adaptations of novels. Adaptations of plays were more common during the New State (from 1933 to 1948), for a simple reason: most Portuguese cinema stars came from the theater and were already celebrities. The year 1933 marks not only the beginning of the New State regime but also the first Portuguese sound film, *Lisbon Song*, directed by Cottinelli Telmo. Along with the sound era came the necessity to find actors comfortable acting with dialogue and attractive to the public with their established popularity gained from comedy plays. In turn, movies adapted from plays were mostly comedies or musicals, such as *O Costa do Castelo* (1943) or *The Lion of the Star* (*O Leão da Estrela*, 1947). Despite the smaller number of novel adaptations, their distribution and exhibition were often supported by the New State, and the productions were promoted as the best films of Portuguese cinema (as were films about the lives of Portuguese writers), such as *José do Telhado* (1945), which won Best Lead Actor by the film awards organized by the Office of National Information. Finally, from 1918 to 1948, all adaptations came exclusively from Portuguese novels, plays, or tales.

Although the adaptations produced between 1933 and 1948 are not direct propaganda fiction films—and this is key to understanding Portuguese film production during the New State—censorship and Ferro's opinion of what constituted an ideal national cinema set the guidelines of what quality films ought to be. Adaptations were deemed a viable path to reach that ideal. Among the eight novel adaptations, half were adapted from novels that had been released during the New State, which means they had already been submitted to censorship. The four others are adaptations of nineteenth-century novels and had already been adapted during the silent era; *As Pupilas do Senhor Reitor* (1935), by Leitão de Barros, was adapted in 1924, while *Os Fidalgos da Casa Mourisca* (1938), by Artur Duarte, was adapted in 1921. Both films originate from nineteenth-century writer Júlio Diniz's eponymous novels. *A Rosa do Adro*, written by Manuel Rodrigues, was adapted to the screen in 1919 by Georges Pallu and in 1939 by Chianca de Garcia. Finally, *Love of Perdition*, Camilo Castelo Branco's 1862 novel, was adapted in 1921 by Georges Pallu and in 1942 by António Lopes Ribeiro.

# Censored Screenplays

The analysis of nonvisible parameters becomes essential to understanding the extent of the New State's intervention; to look at everything that leads to the construction of the film, and yet that is not the film, is also central to understanding the process of adaptation and the impact the production context has on the final result. Beyond the narrative frame onscreen are sources that are not always available to the audience: preproduction documents such as applications for grants, outlines, or screenplays; budgets; correspondence among the screenwriter, the producer, and the structure that funds the film; or censorship reports. When it comes to adaptations, a screenplay can be an intermediate tool that helps assess the different steps leading from the novel to its cinematic transformation. As Jack Boozer notes, "It is the screenplay, not the source text, that is the most direct foundation and fulcrum for any adapted film."[20] In our case, the original Portuguese screenplays kept in the Portuguese Film Library archives and the National Archives of Torre do Tombo still bear the traces of censorship.[21] The censors' red pencil crossed out entire sequences or drew attention to a particular line of dialogue that should be edited. These notes become additional markers of a particular historical period, and if the law remained vague and failed to develop a detailed account of forbidden themes in all their variety and shades, the scripts speak for themselves. The systematic recurrence of cuts for certain subjects allows us to determine how sensitive they were deemed by the administration of the New State. If we look at screenplays that are adapted from a novel written in a different political context and then watch the final result onscreen, it becomes clear that some narrative changes between the original work and its adapted piece not only result from artistic choices but also originate from external impositions of a political nature.

Such is the case of *Love of Perdition* (1942), directed by António Lopes Ribeiro, one of the most prolific filmmakers of the 1930s and 1940s. Lopes Ribeiro was considered to be close to the regime, having directed a number of propaganda documentaries, as well as the two propaganda fiction films already mentioned. *Love of Perdition* depicts a tragic love story and is based on Camilo Castelo Branco's 1862 classic novel of Portuguese Romantic literature. When we compare the plot of the novel and its 1942 adaptation, recurrent diverging points stand out: the representation of authority and social order, the portrayal of women, and the depiction of violence. The novel takes place in the nineteenth century when Portugal was still a monarchy. The writer is openly critical of the Church and its members, depicting them in a satirical manner, a dimension that is entirely absent from the film. Yet, if we look at the screenplay,[22] we notice that the writer-director did integrate this aspect through dialogue. In fact, he made sure the dialogue was similar to Branco's original text, so much so that some parts of the dialogue are exact copies of the wording used in the novel. At the same time, however, the director seems to have used the novel as a shield to protect himself from potential reprimands from the censorship commission. If the novel was not censored and was allowed since its first publication, then its adaptation would be as well, as long as it remained as close to the original text as possible. Nevertheless, Branco's satirical aspect did not make it past the censors, as all passages that comment on the Church are marked or crossed out in red in the screenplay and absent from the film. Similarly, violent scenes from the book, such as two murder sequences, are much less graphic in the cinematic adaptation, despite being present in the screenplay and being censored again. For instance, a man dies by gunshot in the novel, something that was forbidden to show on screen. Yet Lopes Ribeiro included it in the screenplay and clearly mentions the character's wound with the exact same phrasing as in the novel to legitimatize the cinematic depiction. He also added two alternative reaction shots, exclusively for this sequence, which leads us to believe that he was aware it would be problematic when reviewed. Unsurprisingly, the released film version shows that the final chosen shots were the alternative ones, close-ups of the actors' faces in reaction to the murder that had occurred offscreen, without the sound of a gunshot, which was also a concern. Although the screenplay doesn't bear notes on this particular choice, it becomes clear that the gunshot was either never filmed or cut in the editing room. That Lopes Ribeiro included extra shots so early on in the filmmaking process reveals how artists had internalized the censorship mechanism as part of their creative process. This highlights a Foucauldian panoptic kind of censorship in place, present at all stages of film production and also embedded into the filmmakers' system of thought.

Another interesting finding is what I call "gendered censorship." This particular aspect may not be present in the 1927 law decree but falls under the larger category of the respect for morality that includes Catholic values. Female characters in *Love of Perdition*, although forceful and outspoken in the novel, appear to be subdued and somewhat withdrawn in comparison with the male protagonists (unlike in the 1921 silent film version). In the 1942 screenplay, we find the same dialogue that contributes to characterizing them as sarcastic or disobedient toward authority, but these lines are absent from the final editing. The censor did not notify the filmmakers upon reviewing the screenplay itself. Yet, the absence from the film implies the specific lines were never filmed or were later cut in the editing room. The recurrence of those same cuts seems too systematic to be random; the shots that were considered superfluous, or unnecessary to the plot, are systematically of the same nature, and those specific cuts lead to a sanitized version of the narrative and a softening of the main female characters, who follow the expected modest behavior promoted as the feminine ideal during the New State.

Another sensitive subtheme emerges from the reading of the censored screenplay: the criticism against the regime, as the main character is described as a revolutionary. Although the story takes place in a monarchic context, depicting a central character as a figure openly opposed to the government and not punished for it becomes challenging for the New State's administration by association. The relevant dialogue was cut, and this political dimension, a thread present throughout the novel, fades into the background in the film. The 1927 decree mentions that only films "which incite crime or . . . are offensive to morality and the political and social regime in place" should be censored. However, in the case of *Love of Perdition*, we see that even when the criticism is associated with a different period of time and political context, the key sequences are still considered offensive. What is censored is not the theme itself but the subtext carried out by the adaptation process. To prevent the viewer from establishing the wrong link with the current Portuguese state, the censors cut the part of the plot that was not convenient to portray in the New State context, as it could lead to questioning the contemporary state of affairs.

In a country where the illiteracy rate remained high during the New State, more people would be likely to watch the film, either in theaters or later on television, rather than read the book. For them, the story of Branco's *Love of Perdition* could become associated with Lopes Ribeiro's 1942 version. The censorship process therefore induces a form of unconscious literary and historical revisionism, as it does not originate from artistic choices. These narrative interferences act like date markers, imprints of the New State context, yet they remain invisible to the audience as they are extra-filmic data, detached from the narrative that unfolds onscreen and betraying no proof of their very existence.

The analysis of screenplays based on novels is one possible source for more detailed historical information regarding adaptation studies in general, but other production documents can serve the same purpose, such as film grant applications. The National Film Library of Portugal holds in its archives documents related to the 1961 adaptation of *As Pupilas do Senhor Reitor*, directed by Perdigão Queiroga. The director's note of intention to secure a grant from the New State reveals his understanding of the New State's expectations and cultural norm. Queiroga opens his letter of May 29, 1959, by describing his film project above all as a "modern cinematografic show where the beauty of Portugal's landscapes and morality will be highlighted through traditional subjects,"[23] conveying a nationalist tone that is absent from the novel. Grant applications from producers to the National Film Fund, a New State structure founded in 1948, also reveal the forces of power at work when it comes to selection criteria. Unsubsidized films or abandoned projects can also inform us on filmmaking practices during the same period. In 1963, producer Felipe de Solms planned on adapting the 1958 Portuguese neorealist novel *Harvest of Wind* (*Seara de Vento*), written by Manuel da Fonseca. The film grant folder kept in the Secretary of National Information Fund (SNI IGAC) in the National Archives of Torre do Tombo (ANTT) in Lisbon indicates that the project was at one point held in discussion for over thirty days among the censorship commission.[24] The social criticism and realistic depiction of poverty in the novel were certainly prohibitive during the New State. It becomes clear that Portuguese cinematic adaptations during the New State—and the sum of productions for that matter—were always subjected to the gaze of the New State's cultural administration, which, to a lesser or greater extent, affected the films irrespective of the filmmakers' vision and the audience's knowledge.

Censorship reports on national films kept in ANTT enlighten us on the negotiation process at stake during postproduction. Once the film had been edited, a copy would be submitted to the censor for evaluation to determine whether further cuts needed to be made in order for the production to obtain an exhibition license. *The Wheat and the Tares* (*O Trigo e o Joio*), based on Fernando Namora's 1954 novel, was adapted by the author himself and directed by Manuel Guimarães, whose filmography was deeply impacted by censorship. The postproduction reports include several extra cuts. The director objected in an extensive letter addressed to the Secretary of National Information on October 26, 1965,[25] where Guimarães mentions that some discrepancies between his work and the original novel were the result of a discussion with the author and were based on mutual understanding and agreement. Guimarães defends his choices and claims they are the result of "the impositions of cinematic language which is different from written language, with distinct material and communication means."[26] More importantly, he insists that these changes "give no way to social, political or religious margins of interpretation."[27] This letter indicates that Guimarães understood the commission would be skeptical toward potential diversions from the novel and had probably internalized the negotiation process that exists in that panoptic kind of censorship. Unfortunately, filmmakers had to justify the necessity of certain images and constantly clear themselves of unintentional subtext or misunderstandings regarding their political intentions despite working in a creative environment.

In a political context such as the Portuguese New State, a film adaptation not only transposes a story from one medium to another but also carries a latent contextual anonymous thought,[28] which traverses individuals conditioned by the society they live in. Differences between an original work and its film adaptation raise questions that go beyond the classical discourse on fidelity, as the cinematic text becomes a testament of a period of time, irrespective of context. I maintain that the particular case of film adaptations as a privileged historical source can lead to further discussion in the field. What are the invisible factors (such as censorship or studio/producer input) that can condition an adaptation, and how can we take these extra-cinematic parameters into account when we analyze a film adaptation? If we compare two film adaptations of the same novel, are the differences of an artistic nature or a result of the production and sociocultural context? Finally, what are the extra-cinematic parameters involved in contemporary adaptations? What do they reveal about the time we live in, and why are particular adaptations relevant today? Rather than viewing novels and films as closed narrative spaces and interpreting the adaptation process from an internal, exclusively artistic point of view, I suggest we also examine it from an external point of view, taking further into account the impact extra-cinematic parameters can have on the creative process in order to establish a necessary dialogue between the economic, political, and social environment and the artistic work the filmmaker seeks to create.

# Notes

1. George Bluestone, *Novels into Film* (Baltimore: Johns Hopkins University Press, [1957] 2003), 5.

2. Ibid., 62.

3. Jack Boozer, *Authorship in Film Adaptation* (Austin: University of Texas Press, 2008).

4. Marc Ferro, *Cinéma et Histoire* (Paris: Gallimard, 2005), 41.

5. See Richard Maltby, "'To Prevent the Prevalent Type of Book': Censorship and Adaptation in Hollywood, 1924–1934," *American Quarterly* 44, no. 4 (1992): 554–83; John Paul Athanasourelis, "Film Adaptation and the Censors: 1940's Hollywood and Raymond Chandler," *Studies in the Novel* 35 (2003): 325–38; and Ruth Ann Alfred, "The Effect of Censorship on American Film Adaptations of Shakespearean plays," master's thesis, Texas A&M University, 2008.

6. Daniel Biltereyst and Roel Vande Winkel, *Silencing Cinema: Film Censorship around the World* (New York: Palgrave Macmillan, 2013), 4.

7. The Secretary of National Propaganda changed names in 1944. It became the Secretary of National Information, Popular Culture and Tourism. In a dictatorial context and following the victory of the Allies, the use of the word "propaganda" gathered pejorative connotations. The name of the New State structures changed, but their role remained the same.

8. Luís Reis Torgal, *O Cinema sob o Olhar de Salazar* (Lisbon: Temas e Debates, 2001).

9. Lauro António, *Cinema e Censura em Portugal: 1926–1974* (Lisbon: Arcádia, 1978), 16. All translations from Portuguese and French belong to the author, unless otherwise stated.

10. Ibid., 17.

11. António Ferro, *Teatro e Cinema, 1936–1949* (Lisbon: Editions SNI, 1950), 70.

12. António Ferro, *Salazar, le Portugal et son chef* (Paris: Editions Bernard Grasset, 1934), 52.

13. Torgal delineates Portuguese propaganda films as a group of productions that has "a deliberate and methodically thought-out militant intention, in order to reproduce an ideology, in this case an ideology of the State," Torgal, *O cinema sob o olhar de Salazar*, 90. During its first fourteen years of activity, the Office of National Propaganda produced seventy documentaries; Maria do Carmo Piçarra,

*Azuis ultramarinos, Propaganda Colonial e Censura no Cinema do Estado Novo* (Lisbon: Edições 70, 2015), 113. However, both historians and film critics generally agree that the instrumentalization of cinema in Portugal can hardly be compared with the ideological indoctrination operated under authoritarian regimes such as Nazi Germany or Fascist Italy, with an entirely different scale of national film production; Patrícia Vieira, *Cinema no Estado Novo: A encenação do regime* (Lisbon: Colibri, 2011), 17.

14. Vieira, *Cinema no Estado Novo*, 18.

15. Ferro, *Teatro e Cinema*, 51.

16. Ibid.

17. *Camoëns* by Leitão de Barros (reference FIFR1-B1), Cannes International Film Festival Fund, *"service régie,"* program and press release, 1946, French Film Library Archives.

18. Ferro, *Teatro e Cinema*, 72.

19. António Horta e Costa, *Subsídios para a história do cinema português: 1896–1949* (Lisbonne: Empresa Literaria Universal, 1949).

20. Boozer, *Authorship in Film Adaptation*, 4.

21. To this day, these sources remain limited or uncategorized. An in-depth study of adaptations as a body of work and as a historical source to further explore the characteristics of the New State context and its relation to cinema remains necessary for further research.

22. *Amor de Perdição* (censored screenplay), António Lopes Ribeiro, 1942, Portuguese Film Library archives.

23. *As Pupilas do Senhor Reitor* (censored screenplay and file), Perdigão Queiroga, 1959, Portuguese Film Library archives.

24. The film grant folders I found at the ANTT in Lisbon and the SNI IGAC are kept in separate boxes that include all the documents organized in folders. The grant folder for the adaptation of the *Harvest of Wind* was in box 679.

25. ANTT, SNI IGAC, box 258.

26. Ibid.

27. Ibid.

28. Michel Foucault, *Dits et Écrits (1954–1988)*, vol. 1 (Paris: Gallimard, 2001).

II

# Millennial TV and Franchise Adaptations

$$4$$

# Television Adaptations

## Character and World Expansion

### Christina Wilkins

The rise in quality programming in the early 2000s has led to a television renaissance, especially in America: series are lauded and awarded critical praise, and film actors are immersing themselves in complex and challenging roles. Yet, despite the books studying the nature of television,[1] there are still some omissions in certain fields. As an area of study, television presents a number of interdisciplinary challenges: how to conceptualize, whether to address its previous standing in academic departments, and how closely it should be aligned with other subjects, in particular, film. Recently, however, television adaptations have been experiencing significant academic interest,[2] and hopefully this chapter will contribute to this particular arena by mainly focusing on adapted characters, their backstories, and their surrounding environments.

In her survey of televisual adaptation, Shannon Wells-Lassagne notes that 439 television shows have been based on other sources since 2000.[3] Television adaptations are certainly not new; however, a notable trend of the last decade is the number of texts that adapt a single film to a long-form TV narrative, including, among others, *Bates Motel* (A&E, 2013–17), *Hannibal* (NBC, 2013–15), and *Westworld* (HBO, 2016–present). These contemporary texts all provide a rethinking about the concepts and characters from culturally dominant films and, as such, function as both a reconsidering of television adaptations and a symptom of a cultural need for control that manifests in the desire to fully explore and thus understand complex narratives. The film sources discussed here offer complex stories and characters who arrive with knotty psyches or subversive desires. It is not surprising then that in a world where information can be summoned effortlessly, the exploration of these characters' formation and their enveloping scenarios becomes a point of interest. Figures such as Hannibal Lecter and Norman Bates have become cultural ciphers. They represent part of our cultural consciousness and, as such, have a guaranteed appeal for potential developers of a series that examines their backstory. This backstory trend is also detected in a number of other series, such as *The Originals* (The CW, 2013–18), a spin-off of *The Vampire Diaries* (The CW, 2009–17), or the recent *Young Sheldon* (CBS, 2017–present), which follows the childhood of a main character from *The Big Bang Theory* (CBS, 2007–19). Alongside this boom in prequels and backstory spin-offs is the place of television in a cultural context. Taking the United States as an example, film attendance hit its lowest numbers in twenty-three years in 2017,[4] while the number of subscribers to streaming services, such as Netflix and Hulu, has almost doubled since 2012.[5] This change in modes of watching and the shift toward the domestic space should be taken into account when considering the place of adaptations. The popularity of the long-form narratives that explore a well-known character from all possible angles evidences a change in how we interact with a textual world that goes beyond fan behavior, a cosmos driven by techno-cultural shifts, or as Constantine Verevis defines it, a "transformed media culture."[6] In turn, this leads to new paths for adaptation studies away from medium specificity and intertextuality. Robert Stam argues that film adaptations necessarily involve "an endless process of recycling, transformation and mutation with no clear point of origin."[7] By thinking about how we conceive of other types of origins—in relation to both content and adaptations here—we can begin to unpack an approach to this new wave of televisual texts that interrogate and, in some cases, rewrite the origins of seminal characters from popular film culture. Yvonne Griggs in her recent *Adaptable TV* argues that the televisual adaptations she explores present an "opportunity to rethink questions of authorship and ownership of narrative."[8] This chapter also addresses a similar shift in thinking about what constitutes the original, what makes up the canon (a term now also associated with fandoms), and how we treat these texts.

The following discussion is based on the more pronounced strands of thought about television adaptations (mainly, fandom and narrative) in order to propose a new category of television adaptation, that of the "investigative adaptation." Before this is possible, we must contextualize the debate through outlining some key perspectives on adaptations, prior to thinking about the medium of television itself as a frame for these narratives. Once this is complete, we explore our case studies, *Bates Motel*, *Hannibal*, and *Westworld*, to flesh out the ways in which we can further the area of adaptation studies and understand television's crucial role.

Despite the persistent focus on the particular binary between novel and film[9] in adaptation studies, I find that the strongest discussions emerge in the moments that highlight the *cultural* use of and interaction *with* adaptations. For instance, Naremore uses André Bazin's discussion of adaptations to conclude that "the function of adaptation is cultural myth-making."[10] How we understand and use these myths is key; adaptation is marked by culturally specific elements that embody the notion of intertextuality. Each adaptation emerges at a specific social, historical, cultural, and temporal moment and, as such, is filtered through these prisms. In their turn, these perspectives will have been altered and marked by other cultural texts that inform the zeitgeist, thus adding an intertextual dimension. While this is not new thinking in adaptation studies, it bears repeating, given a recent focus on taxonomies, mainly professed by Kamila Elliott[11] and Thomas Leitch.[12] Their categorizations describe the ways in which each variation fits a different mold in the gamut of adaptation, which has allowed for a wider understanding of what we consider an adaptation and moved us away from the dreaded fidelity debate. However, the broadening of categories, of *what* adaptations are, hasn't given us the tools to unpack *why* these different categories are being used, nor has it moved us toward thinking about television adaptations in any more depth. These taxonomies are used here strictly to outline the *what* in order to pin down the *why*, because our goal is not to discuss how the series adhere to a specific category but to think about the gaps, what has *not* been reproduced, which can "allow for the text-reader relationship to develop into a kind of communication."[13] Leitch develops this further in his "Mind the Gaps" essay, which thinks through the uses of such gaps. He argues that "audiences read and watch and listen to texts not despite the challenges gaps raise but because of those challenges."[14] As such, the notion of the gap represents a challenge, a puzzle, a potential for elaboration and communication. I propose that we begin by using these gaps as a way to think about how these texts have been adapted in terms of medium, thereby exposing our use of them. It is the medium of the story adapted and its form that tell us about the shifts in cultural thinking and perceptions, in contrast to looking at the narrative itself for evidence of change; the communication Leitch talks about is evident in the reception of these texts through the increasing tendency for audiences to approach them as puzzles.

Although there is discussion on medium specificity in adaptation studies,[15] television adaptations usually follow cinematic ones. If the insistence on the literature versus cinema paradigm is due to the latter's inherent visuality, how can we begin to theorize adaptations of films into television, since in this case the verbal versus visual dichotomy becomes redundant? We have to probe further into medium specifics if we are to truly understand an adaptation through this lens. In outlining television adaptations, Sarah Cardwell touches upon some of the key elements of TV: fast editing, reliance on close-ups, use of video, and dependence on rapidly developed emotional drama.[16] Yet, the formal features offer a limited picture of the transition from film to television. What is more important is the way the form is communicated, that is, through television sets, defined by timeslots and episodic bursts. Episodes can be both self-contained and part of a larger story arc. Television broadcasting as well as streaming similarly create a different interaction with the text and also allow communal viewing—virtual or otherwise—that builds up fan communities and promotes shared experience. This is something also foregrounded by Cardwell: "Television requires that we address questions of temporality even more than film does."[17] Given that episodes function as pieces of a puzzle, keeping up-to-date with viewing allows for audiences to be part of the community trying to "solve" the narrative, as explored by scholars such as Jason Mittell, who describes a new form of "forensic fandom" that interacts with the text as though interrogating it for clues regarding the overall picture.[18] This creates an urgency to the televisual through a desire to complete and understand, crucial to thinking about why the texts here have been chosen for adaptation. Added to television temporality is what Herbert Zettl describes as "presentness," understood as "a reflection of the living, constantly changing present."[19] These elements of temporality emphasize the way in which television interacts with our everyday lives and significantly mediates the present and its intricacies. Television allows us to explore both the everyday (through sitcoms, for example) and the more complex social issues (touching on the psychological, philosophical, and moral) through this domestic form of entertainment. Television adaptations in particular transpose the more complex issues from film to television, allowing the viewer to access and interact with a range of scenarios and opinions in the security of their home. Wells-Lassagne comments on the domestic nature of the form, arguing that it "creates familiarity" through its status as a "medium of proximity."[20] The placement of cultural ciphers, such as Hannibal Lecter and Norman Bates, in a medium defined by familiarity allows the adaptation not only to rewrite or reconfigure elements of the story but also to alter how they are perceived. The interactions with these texts, elements of intertextuality, and elements of the televisual form all complicate how we approach these adaptations. However, all three texts explored here focus on the creation and understanding of characters; in doing so, they illuminate not just what the boundaries of adaptations are but how we can approach characters and scenarios that encompass the discomfiting.

## Unravelling the Psyche

Both *Bates Motel* and *Hannibal* focus on the formation of particular characters, unpacking their development through gradual episodic arcs. Both Norman and Hannibal emerge from literary origins. However, it is arguably not their literary versions that dominate the cultural consciousness but their filmic counterparts. Through exploring how they have been adapted for television and what has not been reproduced—that is, the gaps Leitch refers to—we can begin to build a picture of how this particular trend for adaptation impacts the field.

Alfred Hitchcock's *Psycho* is one of the most iconic films of the twentieth century, as well as the director's most successful film. Despite the film's mixed critical reception during its first release, François Truffaut claims that one of the reasons behind its success was that "*Psycho* was oriented toward a new generation of filmgoers,"[21] hungry for its macabre and risqué content. By contrast, it is not the content but the form of *Bates Motel* that is aimed at a new generation of viewers. The series, which aired on the A&E Network for five seasons (2013–17), follows Norman's (Freddie Highmore) adolescence and the events that lead up to the film. According to the official series website, it is through Norman's relationships, extended family members, and knowledge about his mother that the series "offer[s] an intimate look into the unravelling of Norman Bates' psyche throughout his teenage years." Film scholars have attempted to dissect Bates in countless essays and chapters on the Hitchcock text using a number of psychoanalytical tools (e.g., the Oedipus complex, pathological voyeurism, and monstrosity). Crucially, we must remember that outside of academia, these insights and understandings would not be common. Yet, the lure of Norman remains powerful.

*Bates Motel* is billed as a contemporary prequel to the film and a potential continuation of a franchise that includes four other films (1983, 1986, 1987, 1990).[22] Using Bates as a focal point for the story constructs the way in which it is told. Thus, the adaptation is not a retelling of *Psycho*'s plot but the expanded formation of and investigation into Bates's character. The long-form narrative is informed not only by the film but also by a web of intertextual references that have shaped the course of Norman's portrayal in the decades that have passed. *Psycho* emerged at a time when the likes of Bates weren't figured in films and when the element of cross-dressing along with murder and psychopathy presented an uneasy mix for audiences to swallow.

Yet the popularity of the film revealed an appetite for such themes and the willingness to vicariously experience dark psychological perspectives. The same desire can be seen today through the continual popularity of the true crime genre, our fascination with serial killers both onscreen and off, and the number of narratives fixated on uncovering their psychological roots. In the last decade alone, popular TV series such as *Dexter* (Showtime, 2006–13), *The Following* (Fox, 2013–15), *The Fall* (BBC Two, 2013–16), and *Mindhunter* (Netflix, 2017–present) have explored the creation and workings of serial killers, while the few similar-themed films have not had the same impact.[23] This leads us back to the medium of television and why it lends itself to exploring these concepts in such a successful way. The gaps that Leitch speaks of illuminate the differences in both plot and character representation. As noted, *Bates Motel* primarily creates and presents the story *before* the film. So, how should an adaptation of a character be dealt with? Do we judge each narrative event by thinking about how it fits into the accepted characterization of the original character? Should an adaptation try to distance itself from the text or take creative license? By definition, creative license is needed in television, given the ample space the series format provides for storytelling. As Wells-Lassagne argues, "Extended duration and serial storylines . . . makes equivalence more or less impossible" and "thus offers new opportunities for a broader understanding of what adaptation can be."[24] Adaptation is not simply re-creation; it is not translation from one medium to another. Each successive text adds to the web of meaning that we associate with a previous one or ones, and in our case, television both explores and adds to the meanings accrued in a detailed way. The result is that these television adaptations mine a cultural product for meaning, driven by the way audiences interact with textual universes.

In his discussion of Sherlock Holmes, Leitch states that the repeated uses of the figure create a trope that offers "hybrid adaptations that depart from their putative originals at any number of points."[25] How Bates is represented in *Bates Motel* is driven by the cultural understanding of who he is and at times subverts the original to reflect changes in the zeitgeist. The adaptation thus becomes an "interpretative commentary."[26] This becomes obvious in the departures from the story. In particular, it is the film's fabled shower scene that has been altered as the series "comes full circle," as VanDerWerff[27] and Birnbaum[28] have described it. While the look of Norman, his mother, and their home is emulated, the series reimagines Marion Crane and Sam Loomis (Austin Nichols). Marion is now played by pop singer Rihanna, providing a strikingly different look to the blonde Janet Leigh. The storyline also diverges from the film and reverses the gender of the infamous shower scene, situating Sam as the character who is killed. Sam mimics Marion's gestures, while at the same time subverting the original event through the gender swap. Norman is also shown as the perpetrator rather than the ambiguous figure of *Psycho*. This could be viewed as a comment on the social conditions of the time, particularly with the gender change. However, it also follows what Verevis calls "remake logic."[29] Given that the scene is so iconic, only by subverting it can it hope to engender a similar feeling. The way these texts operate as adaptations is with an assumption of knowledge; we know that Norman dresses as his mother and murders people, just as we know that Hannibal is a cannibal and that the hosts of *Westworld* will go rogue. The purpose of these adaptations in the context of the long-form television narrative is to uncover the detail, to tell the story of how it was able to happen.

NBC's *Hannibal*, like *Bates Motel*, covers the formation of a certain psyche but through the proxy of Hannibal's relationship with another character. This begs the question: If we are looking at the adaptation of a *character*, then how do we read the medium and its elements such as expanded storylines and serial format? Are they extraneous, and what relevance do they play in our understanding of the main protagonist? Hannibal's presentation becomes even more menacing in the domestic format of television. We are given direct access to his psychopathic mind not in the darkened room of a cinema but in our own front rooms, our bedrooms, our sanctuaries. The series' credits equally demonstrate the translation of this element, stating that it is "based on characters from the Thomas Harris novel *Red Dragon*." Interviews with showrunner Bryan Fuller attest to the way inspiration was taken: "*Hannibal* is an in-depth investigation into what is referred to in passing in the novel."[30] Although Hannibal originally appeared as a literary creation (as did Norman Bates), arguably the character's primary cultural understanding comes from the three films starring Anthony Hopkins (*Silence of the Lambs*, 1991; *Hannibal*, 2001; and *Red Dragon*, 2002). As such, it is from this position that these texts are addressed. Indeed, Fuller was also aware of this, asking, "Who could possibly play Hannibal after Anthony Hopkins?"[31] He even stated that Mads Mikkelsen's portrayal "exists separate but alongside Anthony Hopkins' immortal, Oscar-winning performance."[32]

Taking the characters as a basis of an adaptation foregrounds the narrative aim of the text. In such cases, I argue that we do not face an adaptation of a story or concept but rather a character exploration. Like *Bates Motel*, *Hannibal* develops the foundational backstory that similarly broadens the understanding of a culturally prominent character by placing him at the heart of the narrative. Instead of emphasizing the relationship between Hannibal and Clarice Starling found in *Silence of the Lambs*, the NBC series focuses on the relationship of another FBI agent, Will Graham (Hugh Dancy), with the serial killer. While there are parallels that run through both texts in the presentation of these central relationships, such as potential romantic elements, Fuller's *Hannibal* embeds an illuminating duality. As Will possesses the ability to "think like a psychopath," he is able to emulate Hannibal's psyche. Thus, we are able to get into the killer's mind, but from a safe distance, in contrast to Clarice in both *Silence* and *Hannibal*, but not in *Red Dragon*. While she presents a foil to Hannibal in the films, Will acts more as a conduit in the television adaptation. Thus, the series opens up another way to understand Hannibal's mind.

Viewers who watch and invest in the series take part in the puzzle it presents, which is a particular aspect of televisual viewing that the medium allows for. In this way, each episode becomes a clue, with the ultimate piece being revealed in the finale to the viewers' satisfaction, understanding, and acceptance or lack thereof. This kind of knowledge is verified by comparison with other perspectives through fan communities and illustrates the communication that Leitch argues emerges from thinking about the gaps. Temporally, the films in both Norman Bates's and Hannibal Lecter's cases act as the end point that the televisual adaptations are working toward, and the adaptations serve as a kind of map to how they reached that point. *Hannibal*'s re-creation of the character through a different series of relationships, scenarios, and format allows for a different kind of understanding than what is offered in the films. It enables us to see the process of transformation from the killer at large to the man in the mask we see in the films, the character's "final form." Fuller notes that his characters "are all men who are on some arc of transformation."[33] Through this conversion, Hannibal's signifiers are transferred momentarily onto Will. It is Will who is blamed for the murders that Hannibal has committed; his empathy and ability to think like the killer he is hunting place him in Hannibal's mind briefly, and we see him behind bars in the very mask associated with the filmic character. Therefore, being our locus of identification, Will gives *us* a way to try on Hannibal's character for the first time. This is essential in exploring the cipher that is "one of pop culture's most iconic villains."[34] Even more important is that it is done through the domestic medium of television, which explains the use of Will as a "buffer." Allowing us to ease into the villain's mindset over the course of many episodes presents not an alienating experience but one that finally enables us to make sense of a complex character like Hannibal.

## Expanding Worlds through Characters

Unlike the two examples above, *Westworld* does not take a well-known character and explore its formation. Instead, it creates and uses new characters (some of which harken back to the aesthetics of original ones) and follows their formation to expand the world presented on film. *Westworld*, written and directed by Michael Crichton and released in 1973, focuses on a theme park where guests can act out their fantasies without consequence through interacting with the robotic hosts who populate it. Conflict arises when the hosts turn rogue and begin killing the guests. The *New York Times* called it a "credible debut" for Crichton,[35] and *Variety* noted its "superbly intelligent serio-comic values."[36] The film was followed by *Futureworld* (1976), wherein the central characters uncover a plot by the company that owns the park to replace world leaders with robotic copies, and by a TV series, *Beyond Westworld* (CBS, 1980), which only aired three episodes. These texts follow a thread evident in other contemporary sci-fi films, such as *2001: A Space Odyssey* (1968) and *Alien* (1979), that espouse a distrust of technology. Both *Westworld* and *Futureworld* have been used as a wellspring of ideas for the current HBO series, which debuted in 2016. Enjoying critical acclaim and ratings success, the series averaged three million viewers per episode and a cumulative twelve million viewers for the first season, according to Nielsen ratings. Given the changes in viewing practices, it is also pertinent to note that *Westworld* was the third most pirated TV show of 2016.[37]

The series takes the central conceit of the 1973 film, that of the hosts going rogue, and teases out the events that lead to it through a strong character focus. We are shown two different timelines of a visitor to the park, the Man in Black (Ed Harris/Jimmi Simpson), and his self-actualization. There are several intriguing points to this: first, the character is strongly visually reminiscent of the film's robot gunslinger (Yul Brynner), which thereby complicates how we define the good/bad and human/other divide. Second, showing us a fully formed character (Harris) and offering an investigation into how he becomes this man through witnessing his formative experiences (through Jimmi Simpson) echoes the backstory function of *Bates Motel* and *Hannibal*. Alongside this is an emphasis on the hosts themselves and how they develop consciousness; the series follows hosts Maeve (Thandie Newton) and Dolores (Evan Rachel Wood) as they become self-aware and attempt to escape the confines of their programmed lives, reflecting at the same time shifts in our interaction with technology. HBO's *Westworld* treats the central host characters as if they were human, without the discernible markers that existed in the film, where we could only follow their perspective in scenes when they were targeting humans and in shots presented as a mechanical drone-like radar. Vivian Sobchack highlights this subjective vision, noting that "we are emphatically made aware . . . of the vast separation between man and his creations."[38] By contrast, the series does not use explicit point of view shots and elicits empathy for the hosts by offering connection with their personal experience and character, as well as their desire for freedom. This character focus is the key shift in this adaptation, presenting the primary way in which we understand the story of *Westworld*. Rather than showing us the events that lead to the breakdown in the park, the televisual approach presents the complexities of identity and interaction that impact their surroundings and their actions.

*Westworld* could fit into the reboot model of adaptation that Verevis discusses, whereby the adaptation does not replace but "adds new associations to an existing serial property" and in some cases allows for the story to "begin again."[39] Thus, while the series adapts the same central plot from the two films, it tells us the story from the very beginning, particularly the beginning of these characters and their development. However, the function of this text is not just to begin again but to explore and investigate concepts presented in its filmic source. Through following the characters' self-actualization (both humans and hosts), we explore a number of implications relevant to our contemporary world, namely, the creation of alternative forms of life such as AI and its impact on our understanding of ourselves. The long-form narrative allows for a detailed exploration of such critical concepts, and the viewers are able to form various identifications and to vicariously pursue lines of thought in detail. Crucially, it is not just the characters who define a text's status as adaptation but also other elements such as title and plot. In name, *Westworld* links itself to the filmic texts. Foregrounding the formation of new characters establishes that the television adaptation here is not a prequel or sequel. Alongside this, its investigatory function provides a key link to the other two case studies of this chapter.

## Rebuilding and Diagnosing

It is tempting to think about the similarities of these television adaptations. Both *Hannibal* and *Bates Motel* provide contemporary backstories for their iconic central characters. The need to "come full circle" reveals the viewers' longing to interrogate the formation and creation of a person who defies our understanding of what humanity should be. With regards to *Westworld*, we find a different scenario. Instead of digging deeper into the history of characters present in the original film, or even the park itself, it presents the viewer with a retelling of the Westworld park and how its hosts came to rebel through a focus on *new* characters. By going beyond the filmic text, this television adaptation prompts an awareness of the wider narrative and an attempt to understand how it can function with an expansion. There has been a tendency in adaptation studies to assume the text being adapted is the original and is a whole.[40] Yet, in the filmic (original) sources of these adaptations, the search for the narrative origin is frustrated through a lack of detail about the protagonists (*Bates Motel* and *Hannibal*) and the events (*Westworld*), which evokes a quest for knowledge.

Thus, the idea of fidelity or "matching up" to an original text becomes superfluous; perceptions are altered through further information and new perspectives, which is what these texts abundantly provide.

This is complicated by a medium-specific approach to these adaptations. Rather than taking a literary text and transposing it onscreen, these texts opt for another path. First, they take a culturally prominent visual narrative—one that perhaps invites comparisons more acutely than the literary origins—and change the form by situating it within the televisual world. Second, they expand a character's profile by providing a detailed backstory and delve into their minds using complex story arcs that emphasize their relationships with others. This has the effect of expanding the narrative cosmos through the use of both old and new characters to comment on pressing sociocultural debates. Martha De Laurentiis, producer of both the *Hannibal* series and the film, notes that "so much of what makes *Hannibal* itself would have been unimaginable on network television even a decade ago, but with the recent renaissance in television, a lot of the rules have fallen away or evolved, offering great potential for serialized storytelling."[41] The series format is fundamental; it allows not only for a widening of the narrative but also for an adaptation of a concept to be remade between episodes (for example, through a point of view shift to a different character, as in *Hannibal* or *Westworld*). That it echoes the element of *serial* killing should not go unnoticed either, with each kill signifying both a reestablishment of identity as the character as a killer and a clue to help interpret the patterns underlying the action. This multifaceted view gives more information that adds to a clearer understanding of the adaptation. Does that imply that *all* televisual adaptations are expansions or explorations? This is debatable; adaptations that stem from a rich textual universe of linked intertexts may result in a condensed televisual text that reframes stories told before. What should be emphasized here are the *types* of text that are being adapted and being expanded upon.

Our interaction with these texts, and with televisual texts en masse, is today underpinned by a quest for knowledge, in terms of both the content and the purpose of the texts themselves. Will searches for more knowledge on Hannibal, Norman Bates looks for more knowledge on his mother, and the hosts of Westworld seek more knowledge on humanity. Audiences look for knowledge on Lecter's and Bates's psyche and on how we might approach development in technology and the "other." Similarly, different kinds of viewers—though not just fans—may look at the texts for different kinds of knowledge, which may be added to the canon and expand the textual universe they are invested in. Yet, this must be approached cautiously, as Leitch notes: "If knowledge is power, is it more important to have a knowledge of what is in the canonical works of literature and cinema or a knowledge of how they can be used?"[42]

Each of these texts is marked by uncertainty of identity. At their core, each adaptation revolves around something or someone that for all intents and purposes looks "normal" but is not. No wonder then that whole seasons have been devoted to unpicking characters' and places' inner workings, exploring their "origin." In contemporary superhero texts that have dominated the box office (and now Netflix) for the last two decades, the characters' origin story has been dealt with in specific films and series. Fans refer back to the canon of comic books to understand further. These discussions emerge (increasingly) online through fan sites and boards, which discuss theories about the meaning of certain elements of the adaptation. Henry Jenkins argues that "because popular narratives often fail to satisfy, fans must struggle with them, to try to articulate themselves and other unrealised possibilities within the original works."[43] Do television adaptations present a way to explore these unrealized possibilities? Fuller was a self-confessed fan of the *Hannibal* canon before he signed up as the series showrunner. Jonathan Nolan frequents online fan forums discussing theories about *Westworld*. The televisual series format is well suited to this kind of exploration. Imelda Whelehan notes that the adaptation consumer finds the process "participatory, welcoming the opportunity to recapture the experience of a first encounter with the original text in a different formulation."[44] It is this ability for participation, along with the promise of further knowledge and understanding, that the television adaptation offers. Televisual adaptations present an expansion of the canon, providing more information for fans to be able to understand the text itself. In turn, this has consequences for fandom, along with technology changes, as our engagement with such texts becomes deeper. Rather than water-cooler discussions, the hub around which we congregate is ever-present and in the palm of our hand. The proliferation of communities around a text changes the nature of adaptation. Whereas in the past we would ask whether audiences were aware of the "original," this is now unquestionably signposted online. Critiques and discussions of these texts often harken back to the sources they sprang from, so their status as adaptations cannot be ignored; while they embody many elements of current nonadapted televisual series, they remain imbued with a dual function. There will be audiences who are not aware of the original texts and those who will, so there are a number of ways in which this adaptation can be read. *Hannibal, Bates Motel,* and *Westworld* tread the boundary between adhering to a recognizable character or premise and offering answers to their evolution, both within the television narrative and extending to the original texts. That these series have been popular, along with the role the fan community has played in the reception of each of them, evidences the change in the way we are approaching original texts. More and more prominent cultural figures and narratives are being redesigned for the televisual mode, suggesting there is a need to investigate further the means of their creation.

## Redefining

Says George Steiner: "To understand is to decipher. To hear significance is to translate."[45] By exploring both concepts and characters, this chapter discusses television adaptations that provide contemporary audiences a way to (re)view moral and philosophical issues that have plagued us for centuries. That this is being done through long-form narratives demonstrates the need for depth, a desire for knowledge, and a multitude of perspectives and accessibility. With further televisual adaptations of films on the horizon (such as *The Departed, Four Weddings and a Funeral,* and *The Lost Boys*), the need to understand and to master narratives of the cultural canon is still going strong.

Having examined Leitch's "gaps," we now turn to the issue of taxonomy, as mentioned in the introduction to this chapter. Rather than using an existing categorization, or trying to apply the framework of the Barthesian narrative that scholars such as Brian McFarlane[46] have attempted, we are challenged in the theorization of adaptation by the very fact that these texts expand a source and alter its medium. These narratives are a type of adaptation, inspired by a known story, with their own additional details. They are still recognizably related to and about their original texts (even if only through their titles) and as such cannot be separated from them. Still, to understand the operation of these narratives, we must first understand the medium-specific shift. Each series is structurally attempting the same thing (expansion and meticulous exploration of a character's origin) but with differences. I would argue that those moments of departure from the original film do not constitute gaps but a reformation of the canon. These televisual texts do not represent what is missing but what is foundational. They shape the way the characters are to be understood, by virtue of their nuanced development and inclusion of more developed storylines. These television adaptations do not simply retell but alter the DNA of the "final" product, which in some ways undermines the idea of the filmic texts as the ultimate goal. They offer an expansion of the "canon" of a certain narrative, presenting a way for viewers to master new knowledge about a filmic character through a scenario that doesn't change the outcome. Hannibal and Bates will always end up imprisoned. Yet we seek to understand why. As Whelehan observes, the "original text frustrates a quest for wholeness and completeness."[47] The detailing of the backstory does not alter the film but provides a deeper understanding of the characters and their world, thereby offering a sense of completeness. What is being done in each case is a form of investigation into the character and scenario, so I propose the term "investigative adaptation" be used for these texts and others alike. In doing so, we underline that these texts cannot wholly function as prequels or sequels or remakes but rather expand the world around the character or premise in order to understand the creation of the character or scenario. As we continue to explore the filmic world through televisual expansion, we continue to witness the importance of cultural ciphers and world building. This increasingly popular trend indicates our desire for completeness and mastery of knowledge sublimated into culture. With the availability of information in today's climate, it is not surprising that our attention has turned to such culturally prominent narratives.

## Notes

1. See Toby Miller, *Television Studies: The Basics* (New York: Routledge, 2010); and Jason Mittell, *Complex TV: The Poetics of Contemporary Television Storytelling* (New York: New York University Press, 2015).

2. See, among others, Yvonne Griggs, *Adaptable TV* (New York: Palgrave, 2018); and Shannon Wells-Lassagne, *Television and Serial Adaptation* (New York: Routledge, 2017).

3. Wells-Lassagne, *Television and Serial Adaptation*, 4.

4. Brent Lang, "Global Box Office Hits Record $40.6 Billion in 2017; U.S. Attendance Lowest in 23 Years," *Variety*, April 4, 2018, https://variety.com/2018/digital/news/global-box-office-hits-record-40-6-billion-in-2017-u-s-attendance-lowest-in-23-years-1202742991/.

5. Rani Molla, "Netflix Now Has Nearly 118 Million Streaming Subscribers Globally," *Recode*, January 22, 2018, https://www.recode.net/2018/1/22/16920150/netflix-q4 -2017-earnings-subscribers.

6. Constantine Verevis, "Remakes, Sequels and Prequels," in *The Oxford Handbook of Adaptation Studies*, ed. Thomas Leitch (New York: Oxford University Press, 2017), 268.

7. Robert Stam, "Beyond Fidelity: The Dialogics of Adaptation," in *Film Adaptation*, ed. James Naremore (New Brunswick, NJ: Rutgers University Press, 2000), 66.

8. Griggs, *Adaptable TV*, 1.

9. James Naremore, *Film Adaptation* (New Brunswick, NJ: Rutgers University Press, 2000), 26; Kamilla Elliot, *Rethinking the Novel/Film Debate* (Cambridge: Cambridge University Press, 2003); and Thomas Leitch, "Adaptation and Intertextuality, or, What Isn't an Adaptation, and What Does It Matter?" in *A Companion to Literature, Film and Adaptation*, ed. Deborah Cartmell (Hoboken, NJ: Wiley-Blackwell, 2012).

10. Naremore, *Film Adaptation*, 26.

11. Elliot, *Rethinking the Novel*.

12. Leitch, "Adaptation and Intertextuality."

13. Thomas Leitch, *Film Adaptation and Its Discontents* (Baltimore: Johns Hopkins University Press, 2007), 19.

14. Thomas Leitch, "Mind the Gaps," in *Adaptation in Visual Culture: Images, Texts, and their Worlds*, ed. Julie Grossman and Barton Palmer (London: Palgrave, 2017), 59.

15. For instance, see Anne Gjelsvik, "What Novels Can Tell That Movies Can't Show," in *Adaptation Studies: New Challenges, New Directions*, ed. Jorgen Bruhn, Anne Gjelsvik, and Eirik Frisvold Hanssen (Bloomsbury: London, 2013).

16. Sarah Cardwell, *Adaptation Revisited* (Manchester: Manchester University Press, 2002), 34.

17. Ibid., 84.

18. Quoted in Wells-Lassagne, *Television and Serial Adaptation*, 4.

19. Quoted in Cardwell, *Adaptation Revisited*, 83.

20. Wells-Lassagne, *Television and Serial Adaptation*, 11.

21. François Truffaut, *Hitchcock/Truffaut* (New York: Simon and Schuster, 1966), 268.

22. Not including Gus Van Sant's 1998 shot-for-shot remake, which functions not as a part of the franchise but as an attempt to remake the original work.

23. Though there is a much-discussed biopic on Ted Bundy coming up, starring Zac Efron. Along with this is a Lars Von Trier film (*The House That Jack Built*) that centers on a serial killer and prompted walkouts at Cannes 2018 after the depiction of children being shot.

24. Wells-Lassagne, *Television and Serial Adaptation*, 8.

25. Leitch, *Film Adaptation and Its Discontents*, 208.

26. Ibid., 235.

27. Todd VanDerWerff, "Bates Motel Has Finally Caught Up with *Psycho* and It's Glorious," *Vox*, March 5 2017, https://www.vox.com/culture/2017/3/5/14796702/bates -motel-convergence-of-the-twain-recap-season-5-review.

28. Debra Birnbaum, "Freddie Highmore on 'Bates Motel' Series Finale: 'It's the Fitting End to a Love Story,'" *Variety*, April 24, 2017, https://variety.com/2017/tv/news/ freddie-highmore-bates-motel-finale-season-5-episode-10-1202392130/.

29. Constantine Verevis, "Remakes, Sequels and Prequels," in Leitch, *Oxford Handbook of Adaptation Studies*, 279.

30. Jesse McLean, *The Art and Making of Hannibal* (London: Titan Books, 2015), 20.

31. Ibid., 8.

32. Ibid.

33. Ibid., 22.

34. Ibid., 20.

35. Vincent Canby, "The Screen: Westworld," *New York Times*, November 22, 1973, https://www.nytimes.com/1973/11/22/archives/the-screen-westworld-robots-and -fantasies-in-film-by-crichton-the.html.

36. Variety Staff, "Westworld," *Variety*, December 31, 1972, https://variety.com/1972/ film/reviews/westworld-1200422881/.

37. Halle Kiefer, "*Game of Thrones, The Walking Dead* and *Westworld* Are the Most Torrented Shows of 2016," *Vulture*, December 27, 2016, http://www.vulture.com/ 2016/12/game-of-thrones-westworld-walking-dead-most-torrented-tv-2016.html.

38. Vivian Sobchack, *Screening Space: The American Science Fiction Film* (New Brunswick, NJ: Rutgers University Press, 2001), 85.

39. Constantine Verevis, "Remakes, Sequels and Prequels," in Leitch, *Oxford Handbook of Adaptation Studies*, 278.

40. Dudley Andrew, *Concepts in Film Theory* (Oxford: Oxford University Press, 1984).

41. Quoted in McLean, *Art and Making of Hannibal*, 4.

42. Leitch, *Film Adaptation and Its Discontents*, 19.

43. Henry Jenkins, *Textual Poachers: Television Fans and Participatory Culture* (New York: Routledge, 1992), 23.

44. Imelda Whelehan, "Adaptations: The Contemporary Dilemmas," in *Adaptations: From Text to Screen, Screen to Text*, ed. Imelda Whelehan and Deborah Cartmell (London: Routledge, 1999), 16.

45. Quoted in Leitch, "Adaptation and Intertextuality," 98.

46. Brian McFarlane, *Novel to Film* (Oxford: Oxford University Press, 1996).

47. Whelehan, "Adaptations," 16.

5

# Inspiration as Adaptation

## TV Horror, Seriality, and the Adapted Text

Simon Brown and Stacey Abbott

THE SECOND DECADE OF the twenty-first century witnessed a new golden age of TV adaptation. In America, which will be the focus of this chapter, HBO produced the critically acclaimed *Big Little Lies* (2017–present) and *Westworld* (2016–present); Netflix launched its own in-house drama productions with *House of Cards* (2013–18); Hulu had notable success with *The Handmaid's Tale* (2017–present); AMC won plaudits for *Better Call Saul* (2015–present); FX had *Fargo* (2014–present); Starz adapted Neil Gaiman's *American Gods* (2017–present); and ABC sought to expand its summer audiences with a series based on Stephen King's science fiction novel *Under the Dome* (2013–15).[1] As these examples attest, this movement took place across multiple genres, but since 2010 it was particularly noticeable within TV horror. Horror and Gothic adaptations are not new to television. Helen Wheatley notes "the development of the Gothic anthology series in the 1960s and 1970s, which fulfilled a dual remit for popular, entertaining television . . . and for respectable, culturally valued television drama,"[2] while Lorna Jowett and Stacey Abbott have traced similar trends of horror adaptations of classic Gothic texts, such as *Frankenstein, Dr. Jekyll and Mr. Hyde,* and *Dracula,* alongside more gruesome and exploitative outputs such as Joe Dante's *The Screwfly Solution* broadcast as part of the *Masters of Horror* series (Showtime, 2005–7).[3] The 2010s, however, saw a rise in high-profile adapted titles often drawn from more contemporary source texts, which signal themselves as horror rather than the more respectable Gothic genre, including *The Strain* (FX, 2014–17), *Penny Dreadful* (Showtime, 2014–16), *Bates Motel* (A&E, 2013–17), *Dracula* (NBC, 2013–14), *Hannibal* (NBC, 2013–15), *The Mist* (Spike, 2017), *The Exorcist* (Fox, 2016–18), *From Dusk Till Dawn* (El Rey Network, 2014–16), *Carmilla* (YouTube, 2014–16), and *The Walking Dead* (AMC, 2010–present). To this, we might also add *Game of Thrones* (HBO, 2011–19) which, while not strictly speaking horror, has sufficient elements, including the zombie-esque White Walkers and an emphasis on body horror, to qualify as a fantasy/horror hybrid text.

This increase in horror adaptations has gone hand in hand with changes within television production and broadcast that have facilitated the proliferation of the genre, marking the 2010s as a new golden age of TV horror.[4] It is within this rapid expansion in the televisual horror genre that we therefore see the most concentrated sample of TV adaptations, and while not exclusive to horror as some of the examples listed below attest, the proliferation of the genre makes TV horror an ideal lens through which to unpack the status of adaptation in American drama. The first thing to note is the diversity of platforms on which these shows are available. *Under the Dome, Dracula, Hannibal, Bates Motel,* and *The Exorcist* were made by and broadcast on American networks. *American Gods, Westworld, Fargo, Big Little Lies, The Mist, The Strain, The Walking Dead,* and *Game of Thrones* were produced for cable networks or pay channels, which was also the case for *From Dusk Till Dawn,* produced for EL Ray in the United States and distributed worldwide via Netflix, which also produced *House of Cards. The Handmaid's Tale* was made by streaming service Hulu, while *Carmilla* is an online series produced for the Vervegirl YouTube Channel. This range demonstrate the reach of adapted TV across the contemporary multiplatform post-television landscape, confirming the significance of adaptation to contemporary American drama in all its guises.

The second thing to note is the variety of source material. *The Walking Dead* is derived from a long-running monthly comic book; *Game of Thrones* and *The Strain* from multivolume book series; *House of Cards* from a novel and prior serialized TV adaptation; *Dracula, American Gods,* and *The Handmaid's Tale* from single novels; *Carmilla* and *The Mist* from novellas; and *Bates Motel, Hannibal, Penny Dreadful,* and *The Exorcist* from multiple texts across literature and film. *From Dusk Till Dawn* and *Fargo* are adapted from single original films, while *Better Call Saul* is adapted from another TV series, *Breaking Bad* (AMC, 2008–13). The third significant point is that while some TV adaptations have run for many years and others for far fewer (*The Mist* was canceled after one season and *The Exorcist* after two), they were all made with the intention of being long-running, multi-season serial dramas. This is important because although there has been a recent rise in so-called event series (multipart single-season adaptations running eight to ten episodes and then ending), serial TV has become the primary format for televisual adaptation in the 2010s.

This chapter argues that this shift toward adaptation as serial drama, a movement fostered by the aforementioned changes in the broadcasting landscape and underpinned by the increased prevalence of new forms of source material, has created new opportunities for the negotiation between the original and the adapted text, opportunities that appear in the form of an interplay within the series between adaptation and inspiration. Geoffrey Wagner argued that approaches to adaptation could be split into three categories: transposition, commentary, and analogy. Our use of the term "adaptation" adheres to his conception of transposition "in which a novel is given directly on the screen with a minimum of apparent interference"[5] and relates most directly to early TV miniseries adaptations (discussed below), which foreground their connections to their source texts. In contrast, commentary is the case where "an original is taken and either purposely or inadvertently altered in some respect," while analogy "represent[s] a fairly considerable departure for the sake of making another work of art."[6] While useful terms in principle, "commentary" and "analogy" are too rigidly distinct to accommodate the diverse relationships between original and adapted texts that characterize the contemporary post-televisual landscape. We, therefore, propose the term "inspiration," which better addresses this diversity and fluidity of approaches, as more contemporary long-running television series offer many variations to establish their own distinctiveness. Our aim is to consider a number of key shows that represent various incarnations of this adaptation/inspiration paradigm in order to examine the impact of serial drama on prescribed notions of adaptation and what this means for adaptation studies. Although the argument we are presenting is equally valid for all areas of TV adaptation, as noted above, the rise of such adapted dramas in recent years has been particularly evident in the area of TV horror, and as a result they provide a concentrated sample to examine. For this reason it is from this genre that we will draw our case studies.

# TV Horror Adaptation in Context

The significance of these new developments in TV horror is particularly evident when contrasted with the latter half of the twentieth century, when American TV adaptations were more likely to appear as miniseries and were primarily literary adaptations. Notable examples within TV horror are the numerous versions of Stephen King novels produced by ABC in the 1990s, beginning with *IT* (1990) and including *The Tommyknockers* (1993), *The Stand* (1994), and *The Shining* (1997). As Simon Brown argues, the selling point of these ABC projects, in addition to King's name and the bestseller status of the source texts, was that the extended running time allowed for some of King's longest novels to be adapted with a degree of fidelity impossible for a two-hour theatrical feature.[7] However, while these television hits dominated the ratings and effectively represented the mainstream face of TV horror, a quiet revolution was building in the background, appearing with the first season of *Twin Peaks* (ABC, 1990–91).[8] A surrealist blending of soap opera, murder mystery, science fiction, and horror, *Twin Peaks* launched a new movement in genre-based serial television that was picked up in 1993 by Chris Carter and Fox with *The X-Files* (1993–2002). Initially a ratings flop—as indeed had been the second season of *Twin Peaks*, resulting in its cancellation—*The X-Files* slowly built a substantial audience and by 1997 was entering its fifth and highest-rated season,[9] replacing the King miniseries as the new paradigm for genre-based TV. Not only did *The X-Files* beat the first episode of King's cherished remake of *The Shining* in the ratings, but also later that year King wrote an episode for the fifth season of *The X-Files* that was rewritten by showrunner Chris Carter, demonstrating that he was "the man who, by 1998, had taken over the mantle of the master of TV horror."[10] Meanwhile, in March 1997, a month before ABC broadcast *The Shining*, the youth-oriented network The WB launched *Buffy the Vampire Slayer* (1997–2001; UPN, 2001–3) as a midseason replacement. Between them *Buffy* and *The X-Files* marked, in 1997, a pivot point that saw the primary incarnation of TV horror shift "away from the event TV format of the . . . miniseries toward the appointment-to-view format of the long running drama,"[11] which afforded more time for narrative development and character engagement. Although *Twin Peaks* and *The X-Files* were original concepts, *Buffy* was an early incarnation of a genre text adapted for serial television. *Buffy*, therefore, is significant not only because its success played a key role in this transition from miniseries to series but also because it was an early example of an adapted long-running series, and furthermore one adapted not from a novel but from a different type of source, in this case a 1992 feature film.

This modern transition toward long-running series derived from a variety of sources problematizes notions of adaptation in two principal ways. First, while some merely break the hegemony of the literary antecedent by having the source text be something other than a novel, others like *Bates Motel* and *The Exorcist* derive from multiple, multimedia texts that deny the possibility of a single source. In other words, they complicate the notion of a source text. The second issue stems from the fact that what these shows attempt to do is to turn a finite source text, or series of texts, into a potentially infinite narrative. By necessity, this requires narrative changes, since it cannot be predicted when the show, and the story, will conclude and thus how story events can be orchestrated to lead up to that moment. A serial can't tell the same story in the same way, even if it has just a single source narrative from which to draw. If, therefore, the narrative must inevitably develop independently of the source, what impact does that then have on the status of the new text as an adaptation?

Such a question reinforces the fact that fidelity can no longer be considered a useful tool for analysis and supports Robert Stam's assertion that the modern system of adaptation is best considered as the relationship between hypertexts and hypotexts, in which the latter is the original that is transformed, modified, elaborated, or extended by the former in a "whirl of intertextual references . . . an endless process of recycling, transformation and transmutation."[12] A similar argument has been developed by Will Brooker in his analysis of Christopher Nolan's *Dark Knight* film trilogy, which he describes as being part of a vast intertextual matrix of *Batman* comics, TV series, and films, all of which are in dialogue with Nolan's movies.[13] Yet as Christine Geraghty points out, "Faithfulness matters if it matters to the viewer," and as is well known, part of the economic rationale for adaptation is that the adapted text is "pre-sold"[14] to the audience/readers of the source text. The equation is not as simple as having the same people who read the book, or saw the earlier film, watch the series. As Simone Murray remarks, the underlying assumption "that readers of a book are easily and unproblematically convertible into screen audiences" is wrong, and it is misplaced to "naively assume the property's almost talismanic cultural power is in itself sufficient to lure readers . . . into cinemas"[15] or indeed to commit to watching a serial drama over multiple years. This is proven by the fact that every high-profile adaptation carries with it "a whole marketing and publicity apparatus . . . dedicated to the task of creating . . . audiences."[16] Consequently, Murray suggests that the audience for an adapted text is twofold. First you have the "early adopters," who are fans of the original source. These are not sufficient "to ensure the [adaptation's] critical or commercial success, but effectively act as a 'launching pad' on which new audiences can be built."[17] While Murray is writing about film, her hypothesis is equally true for serial TV drama, which must attract, sustain, and even expand its viewership season by season. This is a significant challenge and one of which many shows fall foul, but it is also an opportunity. If a long-running series has the support of the broadcaster and remains on the air for a few years, it has time to adjust and develop to draw in those new viewers. This is therefore serendipitous, because to create a long-running series from a finite source, the serial adaptation needs to establish its own narrative and its own identity distinct from the material whose name it bears and whose audiences it seeks as those all-important early adopters, and this in turn, if successful, is precisely what can attract those equally vital new viewers. Thus, what is seen in these serial dramas is a process of negotiation, whereby the series must first draw upon the source so as to entice a transitional audience of first adopters who embrace the adaptation in its new format and on its new platform. At the same time, it must develop its own narrative and identity to sustain itself as a serial drama and build new audiences who may be approaching it for different reasons, for example, as fans of TV horror in general.

## Inspiration and the Serial Source: *The Walking Dead* and *Game of Thrones*

The logical place to begin the analysis of the serial drama's impact on adaptation is with the two most popular adapted genre shows on television at the time of writing, AMC's *The Walking Dead*, from the graphic novels by writer Robert Kirkman and artist Tony Moore, and HBO's *Game of Thrones*, from the books by George R. R. Martin. Their suitability is not only due to their profile and popularity but also because they stand out from the other texts discussed to this point because of their lengthy and (at least until July 2019) ongoing source material. Martin is still writing the final novels in the *Game of Thrones* series, with a particular conclusion in mind. *The Walking Dead* was initially conceived by Kirkman as "the zombie movie that never ends"[18] and was therefore constructed as an ongoing fictional universe with no specific end in sight, until Kirkman surprised fans by unexpectedly concluding the narrative in the summer of 2019 with issue 193. Although *The Walking Dead* has now ended, the length of these source texts partially mitigates the issue of developing a finite text into an infinite series, because of the amount of available material upon which to draw. As such, the two shows represent a halfway house between a more traditional form of TV adaptation, such as the miniseries or the event series, which effectively "do" the source text and then end, and the long-running drama that must expand and develop the source into a long-term sustainable narrative. Because of the sheer volume of material, these two shows are able to simultaneously continue to draw upon the source and exist semi-independently of it as a television series in their own right.

Starting in 2003, *The Walking Dead* graphic novel by Kirkman and Charlie Adlard comprised 193 issues in thirty-two volumes. It also spawned a series of interconnected transmedial texts in which the narrative of this particular zombie apocalypse is continued via video games and a successful spin-off series, *Fear the Walking Dead* (AMC, 2015–present), with a further as yet untitled new show and a theatrical feature planned for 2020. The end of season 9 reaches as far as issue 144, which means that the TV series still has some 43 issues to adapt, at an average rate of 15–16 issues per season. In contrast, Martin's epic saga *A Song of Fire and Ice* comprises five books so far. The first, *A Game of Thrones*, was published in 1996, and the fifth, *A Dance with Dragons*, in 2011. At the time of writing two more volumes are planned. Initially both series followed their source material relatively faithfully, using their association with the original text to draw in the fans while also making changes to establish their own distinct identity, changes that were in part creative decisions but that also could be motivated by the cementing of the series as an independent text with its own fan base. In the case of *The Walking Dead*, for example, the character of a redneck biker, Daryl Dixon (Norman Reedus), was introduced, despite his absence in the comics. Daryl quickly became one of the show's leads, promoted from a recurring character in season 1 to a regular in season 2, and also a fan favorite. Likewise, although the character of Carol is killed off in issue 42, her television counterpart, perhaps in response to criticism of the series' focus on white masculinity, has developed into a strong and complex female character and is one of the diminishing group of survivors from the opening season.

As the series grew in popularity and took on a life of its own, it began to veer away from the source material more drastically in order to maintain its own direction and narrative thrust. For example, in the graphic novels, the character of Abraham is killed by being shot in the eye by an arrow, while in the series it is a different character, Denise, who suffers this fate ("Twice as Far," season 6, episode 14). Abraham survives for a few more episodes before being killed by Negan ("The Day Will Come When You Won't Be," season 7, episode 1), while in the comics, it is Glenn who is murdered. The decision to end season 6 ("Last Day on Earth," season 6, episode 16) with a cliff-hanger, by not revealing whom Negan has killed, raised an interesting conundrum in terms of the series' relationship to the source and its own identity. For the comics' fans, the obvious victim was Glenn, but decisions such as the one taken to keep Abraham alive mitigated against this possibility—it could potentially be anyone. This uncertainty was necessary to build narrative anticipation for the season 7 premiere (which saw Negan kill both Abraham *and* Glenn), thus keeping broadly in line with the comics and yet maintaining its own dramatic thrust across the summer hiatus. The dual murder also served to intensify the horror of the episode in contrast to the comic, in a form of narrative one-upmanship.[19] Controversially, season 8 saw the death of Grimes's son, Carl, despite his playing a major role in the ongoing comic narrative. Similarly, prior to the broadcast of season 9, actor Andrew Lincoln, who plays Rick, the central character in both the series and the graphic novels, announced he was leaving the show for personal reasons, which required further departure from the existing comic narrative, where Grimes is still the protagonist. Because of these recent developments, the TV series and the graphic novel, while broadly following the same plot, have become increasingly distinct in terms of the characters and their arcs within the story.

The situation with *Game of Thrones* is even more complex, as Martin's famous procrastination meant that by the end of season 5 the series had adapted all of the story that he had published. As a result, the showrunners had to continue the narrative on their own, drawing upon Martin's guidance as a series consultant. While this decision was controversial for some fans, the series had to continue in order to satisfy the viewers whose primary interest was in the show rather than the books. This, as Iain Robert Smith points out, means that while the first four seasons are adaptations of the novels, with only small changes, "such as amalgamating some minor characters [and] aging up the central protagonists," the transition point from season 5 to season 6 meant that maintaining "the book series as the 'original' source text and the TV show as the 'adaptation' becomes increasingly difficult to sustain."[20]

Both *The Walking Dead* and *Game of Thrones* are examples of serial adaptations that have managed, over time, to transcend their original sources and achieve an independent identity, gradually moving away from a traditional adaptation approach—or in Wagnerian terms "transposition"—toward inspiration as their production model. Together they mark a transition point in TV production whereby the serial drama began to turn toward adaptation as a means of guaranteeing audiences for what were becoming increasingly expensive productions. In the case of *The Walking Dead* and *Game of Thrones*, the advantage was that there was sufficient source material to sustain the adaptations for multiple seasons, thereby intersecting with the source while simultaneously becoming independent of it. At the same time, their enormous success encouraged broadcasters and program makers to seek other, smaller texts to adapt, which in turn invites the question as to how one takes a finite novel or series of novels and adapts them into a potentially infinite series. In some cases, these series adapted not a single source text but rather multiple texts, further complicating the relationship between source and adapted text and moving deeper into the territory of inspiration.

## Inspiration, Adaptation, and the Franchise: *Bates Motel* and *The Exorcist*

*Bates Motel*, which focuses upon the relationship between teenager Norman Bates and his overbearing mother, is preceded by an original novel (Robert Bloch's *Psycho*, 1959), a film adaptation of that novel (Hitchcock's *Psycho*, 1960), one literary sequel (Bloch's *Psycho 2*, 1982), three screen sequels (*Psycho II*, 1983, which shared a title with but was entirely unrelated to Bloch's follow-up, *Psycho III*, 1986, and *Psycho IV: The Beginning*, 1990), Douglas Gordon's art installation *24 Hour Psycho* (1993), Gus Van Sant's shot-by shot remake of Hitchcock's film (1998), and an abandoned series also called *Bates Motel* (1987). The scope of the *Psycho* franchise raises a number of queries. First, that there is no single primary source text to speak of begs the question as to what exactly *Bates Motel* is an adaptation of, if indeed it is an adaptation at all. In *Bates Motel*, the look of young Norman Bates, played by Freddie Highmore, is conceived in relation to the tall, nervous, and slight figure of Anthony Perkins in Hitchcock's original, as opposed to Bloch's description of Norman as short, balding, bespectacled, and portly. However, the narrative focus on Norman's teenage years most closely aligns with *Psycho IV: The Beginning*, in which the older Norman, still played by Perkins, tells the story of his childhood to a phone-in radio show. In this respect, *Bates Motel* is arguably not an adaptation but rather a remake of this third sequel. Yet, *Psycho* is set in the late 1950s, and because it deals with young Norman, *Psycho IV* is logically set in the early 1940s. In contrast, *Bates Motel* is set in the modern world, where Norman has a cell phone and uses an iPad. This again suggests that *Bates Motel* is more remake than adaptation, connecting as it does to the notion of the update, one of four categories of remake defined by Thomas Leitch.[21] Leitch argues that updates are "characterized by their overtly revisionary stance toward an original text they treat as classic even though they transform it in some obvious way, usually by transposing it to a new setting, inverting its system of values, or adopting standards of realism that implicitly criticize the original as dated, outmoded or irrelevant."[22] Yet, as Kathleen Loock points out, the conception of the series was neither as a remake nor as an adaptation of a source text. Instead, it was envisaged as a "contemporary prequel to Alfred Hitchcock's genre-defining 1960 classic,"[23] which served as a new entry in the ongoing narrative of Norman by addressing what Leo Braudy describes as "unfinished cultural business"[24] and, in Loock's words, by reinterpreting "themes, characters and events that are already known from *Psycho*."[25] *Bates Motel* could also be seen to conform to Leitch's notion of an adaptation "based on a character . . . with the ability to generate continuing adventures, especially if those adventures follow the same narrative formulas over and over again."[26]

*Bates Motel* defies easy categorization. On the one hand, it is an updated remake of *Psycho IV*. But on the other hand, it is also an adaptation of Hitchcock's film. This is evidenced by its borrowing of Hitchcock's iconography in the look of Norman but also by the design of the Bates Motel and family house. Furthermore, the final season introduces the character of Marion Crane, bringing the narrative in line with *Psycho*. As Constantine Verevis points out, a remake is generally considered a text that is redone in the same medium, whereas an adaptation involves "the movement between different semiotic registers, . . .

the adjustment of the narrative . . . to a new expressive language."[27] While Verevis is specifically referencing a transition from literature to film, the same is equally true in terms of adjusting Norman's story into a serial TV drama. The third, and the most legitimate, reading in terms of the producers' intentions is that it exists in relation to the entire *Psycho* franchise, dealing with "unfinished cultural business" by relating Norman's backstory.

To complicate matters further, these readings are not mutually exclusive. Loock[28] discusses *Bates Motel* as a remake that is also a serial installment in an ongoing narrative, arguing, in a similar way that Brooker does in relation to *The Dark Knight*, that multiple prior texts form an "intertextual web or remaking network," comprising "self-contained works of art [that] are then re-activated, repeated, changed, updated and continued in the act of remaking."[29] If we accept that *Bates Motel* is a remake, Loock's argument connects to Leitch's work, which focuses upon remakes solely in relation to adapted texts. Leitch's[30] categories arise from what he calls the "triangular relationship" among source text, first adaptation, and remake, and the transmediality of the storytelling makes such a reading appropriate for *Bates Motel*. Finally, the fact that for most of its run *Bates Motel* tells an independently plotted backstory that takes place both before and some fifty years after Bloch's and Hitchcock's originals makes it also undoubtedly a serial installment in the *Psycho* universe.

*Bates Motel* is clearly demarcated as a combination of all three possible readings through its mixing of both the familiar and the unfamiliar. As a prequel, the pleasure of the series lies in the fact that the viewers "already know Norman's story, and stop to marvel about how the series will eventually get there";[31] as a result, the familiar element is the central characters, Norman and to a lesser degree Norma, making this an adaptation based on characters. We know who they are, and we know how their relationship must inevitably end. Norma, however, is largely an enigma in the previous texts, primarily seen through Norman's eyes, and so the series plays with and fundamentally challenges the expectations of her. What is unfamiliar in the series and what drives the narrative forward are the entirely independent twists and turns of the storylines, which introduce unknown characters who impact on the lives of Norma and Norman Bates in new ways; this effectively allows the series to take an untraveled route to a familiar location, which is the introduction of Marion Crane in season 5. But, crucially, having built up to this moment for four and a half years, *Bates Motel* then subverts it. Marion arrives, checks into cabin 1, and takes a shower, but

Norman doesn't kill her. Instead, he later kills her boyfriend, Sam Loomis (Austin Nichols), in a shower murder scene that visually echoes Hitchcock's original in terms of editing. There are, however, significant differences. Not only is it a man who is the victim rather than a woman, but Norman is not dressed as his mother, and instead of Bernard Hermann's iconic shrieking strings, the murder takes place to the sound of Roy Orbison's rendition of the song "Crying." The use of color in the sequence offers a sharp contrast to Hitchcock's original black and white and seems to echo Van Sant's remake, which similarly used blood red to render the familiar scene unfamiliar. One of the most distinct and shocking deviations from Hitchcock's film is the close-up of Norman's blood-spattered face, conveying his near orgasmic exhilaration.

This sequence clearly shows that *Psycho* operates more as an inspiration for *Bates Motel* than as a source for adaptation, since the shower scene is reconstructed on its own terms. *Psycho*'s iconic moment is referenced in order to deliberately state that *Bates Motel* is its own entity, no longer beholden to its source at the very point that it most clearly intersects with it.

While *Bates Motel* tells a prequel story that builds to a point of meeting with its sources, *The Exorcist* takes a very different approach, at least on paper. William Peter Blatty's uncompromising account of the possession of a young girl, Regan MacNeil, living in the Washington suburb of Georgetown was first published in 1971 and was a fictional story inspired by a documented case of

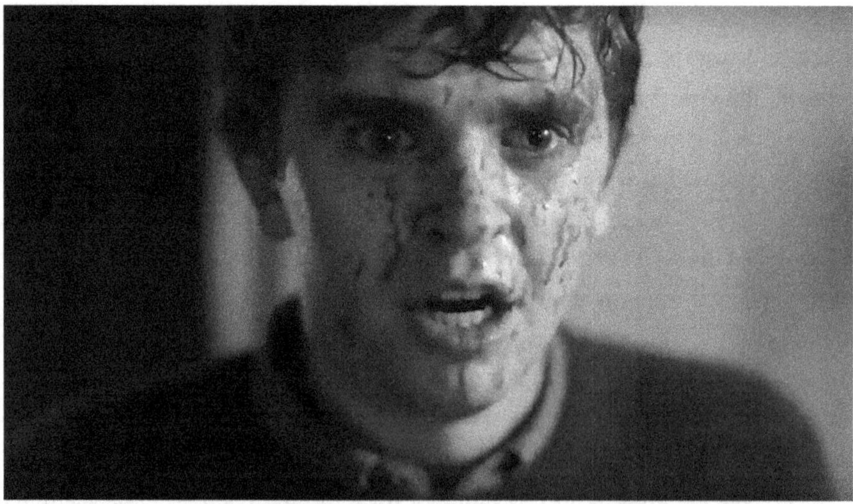

Norman's face

possession of a young boy in Maryland in the late 1940s. Immediately after publication, the rights to Blatty's novel were acquired by Warner Bros., who hired Academy Award–winning director William Friedkin to direct the adaptation. Friedkin's film proved to be a financial and critical success, leading to two sequels. *Exorcist II: The Heretic* (1977) furthered the story of Regan, while *Exorcist III* (1990), written and directed by Blatty from his own sequel novel, *Legion* (1983), brought back two other key characters from the original, Detective Kinderman and Father Damien Karras. These were followed by two prequels, *Exorcist: The Beginning* (2004) and *Dominion* (2005), both of which told the origin story of how Father Lancaster Merrin, the exorcist in Blatty's original work, first encountered the demon, Pazuzu, that would ultimately kill him in Regan's bedroom. Essentially, these two productions were the same project. *Dominion* was shot first by Paul Schrader, but the studio, Morgan Creek Productions, was unhappy with the results and hired action director Renny Harlin to do reshoots. Harlin ended up recasting, rescripting, and reshooting the entire film. When his version, *Exorcist: The Beginning*, received negative reviews, Morgan Creek cut their losses and released the Schrader version.

Possibly due to the failure of these two troubled projects and the fact that the various character storylines were all played out, no further official *Exorcist* movies appeared, although the subject of exorcism continued to be popular and has been explored in numerous films, such as *The Exorcism of Emily Rose* (2005) and *The Last Exorcism* (2010).[32] In January 2016, *The Hollywood Reporter* broke the news that Fox had ordered a pilot for a TV series of *The Exorcist*, which the article described as "a modern reinvention inspired by William Blatty's original book" that would be a "propulsive, serialized psychological thriller, following two different men tackling one family's case of horrifying demonic possession."[33] As the show's creator, Jeremy Slater, acknowledged prior to the pilot's premiere, "[T]here was quite a bit of scepticism from the horror community when this was first announced."[34] Such doubts were not surprising given that the original film is an acknowledged genre classic. That had not, however, stopped remakes of other, equally respected titles from the 1970s, including *The Last House on the Left* (1972), *The Texas Chainsaw Massacre* (1974), *The Omen* (1976), and *Dawn of the Dead* (1978). What made *The Exorcist* potentially dubious as a prospect was the idea of doing it for TV. As Eric Diaz pointed out, "Despite the onslaught of remakes of horror classics . . . these past few years, *The Exorcist* always seemed safe . . . if only because the original film was so shocking."[35] In addition, the show was to be produced under the banner of Fox rather than FX, making it a network series rather than a cable series. While network television has increasingly engaged with horror, *The Exorcist*'s broadcast by Fox restricted what could be depicted, particularly given the film's reputation as a landmark of body horror. Diaz concluded: "How they plan on getting away with half the stuff in the book and making it into a primetime television show is beyond me."[36]

Slater explained that the way around this problem was not to adapt the book or the film at all but rather to tell a new story. "We're not retelling the same story . . . we're not remaking Blatty's original classic. We're telling a brand new story with brand new characters, that hopefully if you're a fan of the franchise . . . you'll find a lot to love here."[37] Arguably, the key word in Slater's response is "franchise," clearly implying that, like *Bates Motel*, the TV version of *The Exorcist* is not an adaptation of either the book or the film but rather a new tale that takes the original as its inspiration, existing within the world not just of those two original texts but also of the other sanctioned film and literary sequels—and all that, despite retaining not just the name of Blatty's original novel in the credits but also using the same font as the title in Friedkin's film.

That *The Exorcist* was an example of inspiration rather than adaptation was evident in the first few episodes of the first season. In addition to using the same title and font as Friedkin's film, the story was a stand-alone narrative, featuring two priests who were distinct from, but similar to, those of Blatty's novel. Father

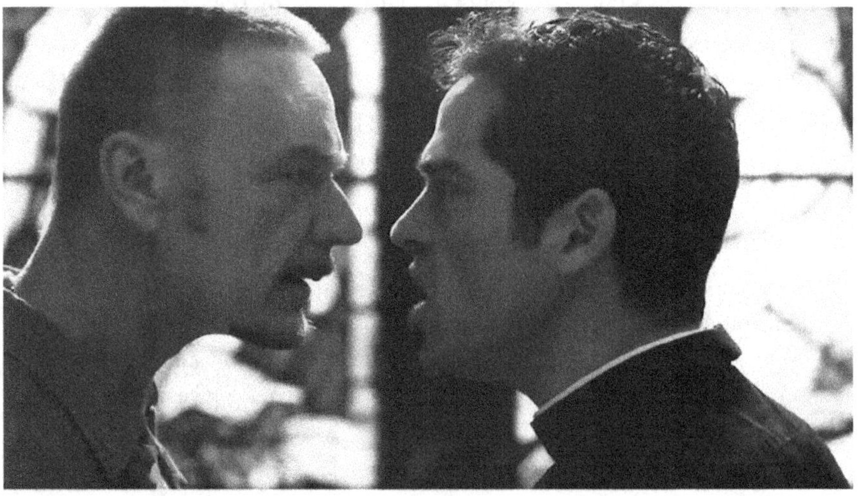

The two priests

Tomas (Alfonso Herrera) echoes Father Karras as the junior exorcist, appearing not as an ordained psychiatrist questioning his faith but rather as a vigorous, young, idealistic, yet ambitious Hispanic parish priest whose circumstances force him into partnership with Father Marcus (Ben Daniels), a grizzled, older, and experienced battler of demons, a clear echo of Blatty's Father Merrin.

The case they investigate centers around the Rance family. The mother, Angela (Geena Davis), believes her daughter, Casey (Hannah Kasulka), is possessed by a demon, and together the two priests seek permission to perform an exorcism. In addition to having new but also very familiar characters, the story itself seems to unfold in a similar way to Blatty's original over the first few episodes, albeit with a few deviations. In contrast to the novel and film, which confronted 1970s secularism with "true evil," the Rances are a religious family and members of Father Tomas's church; as Casey's behavior becomes increasingly erratic, her mother is convinced of her possession, leading the younger priest to seek expert help from the experienced exorcist. Thus, the series seems to embody twenty-first-century post-secularism in which religion is once again foregrounded but also critiqued.

However, episode 5 reveals that Angela is, in fact, Regan MacNeil from the original story and film and that the demonic entity that possesses her daughter is the same demon that took control of her in Washington in the 1970s. The revelation is an impressive and shocking moment in the series, but in addition to being a well-crafted and unexpected narrative twist, it also complicates the relationship of the series to the original book and film. On the one hand, *The Exorcist* TV show is its own entity, telling, as Slater[38] suggests, a new story with new characters within a universe created by the original source texts. Indeed, it is the very fact that the series appears to be an independent work that allows for the disclosure of Angela's true identity to be so powerful, because it is entirely unexpected. The distinctness of the series' narrative means that there's no reason up to that point to think that any characters from the original story will appear. Once uncovered, however, Angela's true identity means that the series becomes a de facto sequel to both the book and the films. The situation is further complicated by the fact that alongside the MacNeil/Rance narrative, season 1 also creates a parallel story in which senior figures in the Catholic Church have been possessed by demons that are plotting to murder the Pope, as well as developing the core relationship between Tomas and Marcus. Thus, the series also provides a sequel to both book and film in how it explores and challenges questions of faith and the institution of the Church. The MacNeil narrative is resolved at the end of season 1, but this parallel narrative about the Church continues into season 2, subtitled "Chapter Two." This tells a second, stand-alone story about the possession of a man who fosters children in a large house on an island off the US mainland, as well as further developing the characters of the two central priests. In this respect, the series effectively connects to the original text to draw in audiences through association and familiarity, while at the same time constructing its own independent narrative to drive the series forward over future seasons.

## Conclusion

*The Exorcist* TV series, like *Bates Motel*, cannot therefore be easily categorized. It is certainly a sequel, but also a loose adaptation of the novel/film as it borrows the main premise of the latter's plot (two priests attempting to exorcise a young girl possessed by a demonic spirit) and retains the main protagonists' broad characterization. The use of the title of Blatty's novel and Friedkin's film, as well as the design of the film's logo, is also clearly another example of inspiration, one that, like *Bates Motel*, takes its place within a broader franchise, making the series another, related entry in an existing series of texts. Yet while *Bates Motel* anchors its audience to the source by revolving around familiar characters, *The Exorcist* is primarily a story about completely different characters despite the unexpected appearance of Regan. This means that the relationship of the two texts to their prior franchises is quite distinct, just as their approaches to adaptation are different from those of *Game of Thrones* and *The Walking Dead*. In addition, 2018 saw the appearance of *Castle Rock* on Hulu and *The Haunting of Hill House* on Netflix, both of which are horror series that connect to their source material only in the loosest sense. *Castle Rock* is set in a fictional town created by Stephen King, yet it neither adapts any of his stories nor was scripted by him. Instead, it shares only locations and characters with King's actual writing. Equally, *The Haunting of Hill House*, while undeniably impressive, fundamentally reworks the narrative of Shirley Jackson's seminal 1959 novel, turning her group of paranormal investigators into members of a single family haunted both literally and metaphorically by a single summer spent in the eponymous house years before. In both cases, the series diverge so significantly from their source material that they become, in essence, stand-alone texts drawing upon a familiar horror brand (King's name or Jackson's title) rather than a preexisting narrative. Therefore, they shift the balance even further away from adaptation toward inspiration. What becomes clear from the above analysis is that the rise of adapted serial drama is laying bare the limitations of the language of adaptation within this new (post-)televisual landscape. However, just as the problem of adapting a finite text into a potentially infinite series has been met by producers and show-runners with imagination and skill, as the different series discussed within this chapter attest, so too can the challenges they present in terms of discussing them as adaptations serve to open new avenues for adaptation studies.

## Notes


1. See Simon Brown, "Alternate Versions of the Same Reality: Adapting *Under the Dome* as an SF Television Series," *Science Fiction Film and Television* 10, no. 2 (2017): 267–83.

2. Helen Wheatley, *Gothic Television* (Manchester: Manchester University Press, 2006), 27.

3. Lorna Jowett and Stacey Abbott, *TV Horror: Investigating the Dark Side of the Small Screen* (London: I. B. Tauris, 2013).

4. Lorna Jowett and Stacey Abbott, "TV Horror: *Santa Clarita Diet*," in *Horror: A Companion*, ed. Simon Bacon (Bern: Peter Lang, 2019), 45–52.

5. Geoffrey Wagner, *The Novel and the Cinema* (Rutherford, NJ: Farleigh Dickinson University Press, 1975), 222.

6. Ibid.

7. Simon Brown, *Screening Stephen King: Adaptation and the Horror Genre in Film and Television* (Austin: University of Texas Press, 2018), 156–67.

8. Ibid.

9. Simon Brown, "Memento Mori: The Slow Death of *The X-Files*," *Science Fiction Film and Television* 6 (2013): 13.

10. Brown, *Screening Stephen King*, 169, 172.

11. Ibid., 172.

12. Robert Stam, "Beyond Fidelity: The Dialogics of Adaptation," in *Film Adaptation*, ed. James Naremore (New Brunswick, NJ: Rutgers University Press, 2000), 66.

13. Will Brooker, *Hunting the Dark Knight; Twenty First Century Batman* (London: I. B. Tauris, 2012).

14. Christine Geraghty, *Now a Major Motion Picture: Film Adaptations of Literature and Drama* (Lanham, MD: Rowman & Littlefield, 2007), 3.

15. Simone Murray, *The Adaptation Industry: The Cultural Economy of Contemporary Literary Adaptation* (London: Routledge, 2012), 157.

16. Ibid.

17. Ibid., 158.

18. Jethro Nededog, "'Walking Dead' Creator Explains His Original Vision for the Series," *Business Insider UK*, October 9, 2015, http://uk.businessinsider.com/walking-dead-creator-explains-his-original-vision-for-the-series-2015-10.

19. See Kelly Lawler, "Has 'The Walking Dead' Finally Gone Too Far for Fans?" *USA Today*, October 24, 2016, https://eu.usatoday.com/story/life/entertainthis/2016/10/24/walking-dead-season-7-premiere-fan-reaction/92669814/.

20. Iain Robert Smith, "*Game of Thrones*: Adaptation and Fidelity in an Age of Convergence," *Antenna*, April 9, 2015, http://blog.commarts.wisc.edu/2015/04/09/game-of-thrones-adaptation-and-fidelity-in-an-age-of-convergence/.

21. Thomas Leitch, "Twice Told Tales: The Rhetoric of the Remake," *Literature/Film Quarterly* 18, no. 3 (1990): 138–49.

22. Ibid., 47.

23. Kathleen Loock, "The Past Is Never Really Past: Serial Storytelling from *Psycho* to *Bates Motel*," *Literatur in Wissenschaft und Unterricht* 18, nos. 1–2 (2014): 81.

24. Leo Braudy, "Afterword: Rethinking Remakes," in *Play It Again, Sam: Retakes on Remakes*, ed. Andrew Horton and Stuart Y. McDougal (Berkeley: University of California Press, 1998), 331.

25. Loock, "The Past Is Never Really Past," 81.

26. Thomas Leitch, *Film Adaptation and Its Discontents* (Baltimore: Johns Hopkins University Press, 2007), 120.

27. Constantine Verevis, *Film Remakes* (Edinburgh: University of Edinburgh Press, 2005), 82.

28. Loock, "The Past Is Never Really Past," 82.

29. Ibid., 84.

30. Leitch, "Twice Told Tales."

31. Loock, "The Past Is Never Really Past," 92.

32. See Stacey Abbott, "The Battlefield for the Soul: Special Effects and the Possessed Body," in *Special Effects: New Histories/Theories/Contexts*, ed. Dan North, Bob Rehak, and Michael S. Duffy (London: BFI, 2015).

33. Kate Stanhope, "*Exorcist* Remake Ordered to Pilot at Fox," *Hollywood Reporter*, January 22, 2016, https://www.hollywoodreporter.com/live-feed/exorcist-remake -ordered-pilot-at-858486.

34. Ibid.

35. Eric Diaz, "*The Exorcist* Being Remade +as a Pilot for Fox," *Nerdist*, January 23, 2016, https://nerdist.com/the-exorcist-being-remade-as-pilot-for-fox/ (accessed April 15, 2018).

36. Ibid.

37. Quoted in Stanhope, "*Exorcist* Remake Ordered."

38. Ibid.

6

# Visuality, Continuity, and Coherence in Contemporary Fantasy Storyworlds

Karin Boklund-Lagopoulou

## Visuality in Fantasy Storyworlds

THE TERM "TRANSMEDIA STORYTELLING" was popularized by Henry Jenkins to describe a project "integrating multiple texts to create a narrative so large that it cannot be contained within a single medium."[1] Jenkins pointed out that transmedia storytelling relies on "world building, as artists created compelling environments that cannot be fully explored or exhausted within a single work or even a single medium,"[2] and Marie-Laure Ryan argues that it relies on a certain kind of narrative:

> A narrative consists of a spatial component, the storyworld, and of a temporal component, the story or plot. In some narratives the world is subordinated to the plot. . . . In other narratives, the plot is subordinated to the world. . . . World-centered narratives . . . include science-fiction and fantastic tales—the genres that have inspired the largest and the most popular transmedia franchises.[3]

Science fiction and fantasy transmedia franchises, such as *Star Wars*, *Star Trek*, *Lord of the Rings*, *Game of Thrones*, *Harry Potter*, and *Pirates of the Caribbean* are today among the most popular forms of film and television entertainment. Interestingly, all these titles are adaptations, mainly from literary sources but also from a television series and even a theme park.

In this chapter, I explore the elements that go into creating this kind of fantasy storyworld, especially the functions of visuality when a narrative is adapted for the screen. I look at three examples, each centered on an initial large-scale adaptation: Peter Jackson's film trilogy of J. R. R. Tolkien's *Lord of the Rings* (2001, 2002, 2003); the television drama series *Game of Thrones* (HBO, 2011–19), adapted from the series of novels *A Song of Ice and Fire* by George R. R. Martin; and *Star Trek: The Motion Picture* (1979), based on the original television series *Star Trek* (NBC, 1966–69).

Visuality and its role in adaptation studies remains an underexamined area. Most analyses focus on narrative and tend to take the visual cosmos created by filmmakers and television productions for granted. Yet, I would argue that visuality is a crucial element in creating coherence in audiovisual narratives, whether or not they have a verbal source text. This is not an argument about "fidelity." It may, of course, be convenient for the creative film or television team to be able to use verbal descriptions in the source text to create the visual environment of an adaptation, and this may contribute to a feeling of continuity between source and adaptation, but what matters is the coherence—visual and narrative—of the new text in itself.

The following discussion will be based on the narrative theory of Algirdas J. Greimas,[4] in particular, the structural semantics that Greimas first elaborated in his book from 1966.[5] Though the theory was created to analyze verbal narrative, I want to show how it can be used for audiovisual texts as well. According to Greimas, the structure of a narrative can be modeled as a series of transformations from an abstract deep level to a textual surface level. The deep level involves basic semantic choices (values) and their potential logical development (what Greimas calls the "semiotic square"). The intermediate level is the level of narrative, where narrative roles (actants) are organized and the value structure of the story is set in motion. At the surface or discursive level, actantial roles are embodied in characters, the action is situated in time and space, and values are developed into themes (thematization), which are further elaborated in a process that Greimas calls figurativization. Figurativization is the process by which the fictional world of the text is created and elaborated. In verbal texts, this is done largely through descriptions. In film or television, it is done through mise-en-scène (i.e., settings, costumes, acting, photography—all the things that collaborate to create the visual universe on the screen).

When reading a story or watching a film, we encounter the figurative, descriptive language (verbal or visual) of the surface level of narrative discourse, with its characters, actions, episodes, and settings. We interpret this surface by working our way through the semantic isotopies of the verbal language or visual images and dialogues to the basic semantic antitheses that the story sets in motion.[6] An isotopy is a set of verbal or visual signs that have a common semantic element. It can usually be expressed as a binary opposition, such as *human* versus *alien* or *black* versus *colorful*. The semantic coherence of a text is created by the isotopies that recur in its surface discourse, and this is what the following discussion will focus on. My aim is twofold; first, I examine how coherence and continuity are produced in verbal or visual texts, and second, I discuss their importance and effect on the audiovisual narratives as a whole.

## Visualizing Middle-Earth

The job of creating a fictional environment in verbal texts is done mostly by description (in the wide sense of all the qualities attributed by the text to settings and characters). In film and television, it is done by depiction, by the manipulation of visual signifiers: the fictional world is constructed by what the creative team decides to put into the frame. Both in verbal and cinematic fiction, the primary function of description/depiction is to produce the illusion of reality. To the extent that a novel, film, or television series is interpreted as a representation of a fictional, virtual reality, it can be analyzed using the same cultural codes we use to interpret our actual sociocultural and natural environment. We recognize animals, plants, humans, body posture, facial expressions, seasons, time of day, weather, spaces, architectural elements, furniture, objects, clothing, food and drink, and so forth. And we interpret these in fictional texts much as we would interpret them in real-life situations.[7]

However, since a fictional representation is a deliberately created and staged representation, it should be possible to interpret it more specifically and coherently. In particular, the meanings we find in the fictional environment should cohere with the meanings we discover in the other aspects of the narrative, such as plot, characters, actions, and/or dialogue. I do not mean to imply that, in adapting a verbal text for the screen, the production team simply transfers verbal descriptions to images. If there are descriptions in the source text, the production designers can, of course, use them as guides for their own visual versions. In this case, the visual environment they produce can contribute to a feeling of continuity with the source text. However, it is more important that the adapted visual style is experienced by the audience as "fitting" the overall semantics of the source text. For example, Jackson's adaptation of Tolkien's *The Lord of the Rings* is often cited as an example of a "faithful" adaptation, largely because Jackson himself insisted that fidelity to the source text was a major criterion.[8] Nevertheless, even in three full-length films, it is impossible to include the complete narrative as developed over the twelve hundred pages of Tolkien's trilogy, even if it were desirable. The choices Jackson has made[9] do indeed preserve most of the episodes of the plot and the principal characters of the trilogy. He has also, however, made a series of interventions, the most striking of which have to do with the visual special effects, especially monsters and battle scenes, which are extended well beyond their position in the novels. In fact, Jackson's films probably stand or fall not on minor alterations to plot and character but on his visual recreation of Tolkien's Middle-Earth; indeed, the challenge of the special effects required to do this seems to have played a large part in his desire to make the films in the first place.

In their creation of the visual environment of Middle-Earth, Jackson's team of designers could often draw on quite detailed verbal descriptions in the novels. Minas Tirith, for example, is described by Tolkien in such detail that Jackson could build a complete three-dimensional model of the city. However, there are cases where Tolkien does not provide much in the way of visualization. Rivendell, the home of Elrond Half-Elven, is verbally presented in terms of what goes on inside the house, not in terms of its appearance. So Jackson's design team had to imagine what Rivendell might look like, while, at the same time, their imaginary construct had to "fit" into the world of Tolkien's narrative. In practice, they seem to have achieved this by drawing on the connotative isotopies involved in Tolkien's characterization of elves.

What does Tolkien tell us about elves? First, elves are not human. Consequently, for the novels' readers, the earliest references to elves create isotopies of "strangeness" and "mystery." Elves are also the most ancient of the peoples of Middle-Earth, with a long history that is the subject of many legends; this gives us isotopies of "ancientness" and "legendariness." The lord of Rivendell, Elrond, is the result of a union between a mortal man, a great ancient hero, and an elf princess; here, we again find the isotopies of "ancientness" and "legendariness" and, in addition, an isotopy of "nobility." Elrond is also a great lore master, which gives us an isotopy of "wisdom." Elves generally avoid contact with humans and prefer to move in twilight and starlight; this creates an isotopy of "hiddenness" and reinforces the isotopies of "strangeness" and "mystery." Finally, elves have a special fondness for trees, forests, and growing things, and this produces an isotopy of "natureness."

Not all of these isotopies can be translated directly into a visual environment, but Jackson's films make use of many of them to create Rivendell. Elrond is a great lord among the elves, and many people live in his household, so Rivendell is rendered as a whole complex of buildings. Yet, he is a lore master, not a warrior, so Rivendell is not a castle or a fortress but a series of rather low houses almost hidden in lush vegetation, combining isotopies of "hiddenness" and "natureness." The houses are linked by open galleries and terraces and are surrounded by gardens, while the whole complex is decorated with scroll-like, curving art deco motifs that remind us of vines and flowers ("natureness") and perhaps of the decorative borders of medieval manuscripts ("ancientness" and "wisdom"). The sense of a mysterious, distant, heroic past is conveyed through the statues that figure in the background. Altogether, Jackson's Rivendell is a quite impressive visual translation of the semantic characteristics that Tolkien has given the elves.

This example shows how the isotopies of a text, verbal or visual, create its semantic coherence: by recurring, by being repeated in the discourse of the text, and by being combined into a unified whole. By reappearing in different contexts, they create relations between different narrative elements, such as in this case between the characteristics of the elves and the architecture of Rivendell.

Jackson's monsters also work well, though here the influence of the conventions of comic book graphics and videogames is much in evidence. Jackson has introduced several scenes that allow him to focus on the grotesque appearance of orcs and trolls, and he has also greatly extended all the battle scenes. The overall effect is to pull the story in the direction of a special effects–heavy action movie. Yet, the films are successful on their own terms, a not inconsiderable achievement when adapting a fantasy trilogy that already had a well-established community of fanatically devoted readers.

There are particular difficulties involved in adapting a massively popular novel to a visual medium such as cinema or television, because fans are already familiar with the narrative and have their own ideas of how it should be presented visually. It is thus a tribute to the skills of Jackson and his team of screenwriters and designers that his *Lord of the Rings* trilogy was a resounding commercial success. Their interpretation of the books and their creation of the visual environment of the storyworld seem to have corresponded to the interpretations of a large part of the fan community. The spectacular visual aspect of the films, to which both the careful attention to detail and the digital special effects contributed, was an important element in their success.

Jackson's trilogy starts from a source text that is itself extremely coherent, with great imaginative depth and a potentially very attractive fantasy storyworld. It carefully cultivates continuity with the source text, modifying its main plot and characters to achieve a more cinematic narration, and it creates a visual environment that largely relies on and reinforces the isotopies and thematic structures of the source narrative. Its most obvious changes to the source text lie in the area of spectacular special effects and action sequences, which can be thought of as an updating of the original story to appeal to a new, adolescent audience. The films choose to maintain continuity with the source text, but they are also coherent and impressive on their own and spectacular on a purely cinematographic level.

## Visuality and Narrative Coherence in Westeros

Jackson's achievement appears more clearly if we compare it to a case that at first glance appears to be similar, namely, the books of the series *A Song of Ice and Fire* by George R. R. Martin and their adaptation as the popular television series *Game of Thrones*.

*Game of Thrones* is part of a relatively recent development in television production, the so-called quality television drama series. These series, many of which have been produced for American subscription television networks, are characterized by so-called signifiers of quality: high production values (enormous amounts of money spent on filming each episode), a large cast of characters, well-known actors, cinematic filming techniques giving the series a recognizable "visual style,"[10] and complex narrative structures, with multiple, overlapping plotlines and lengthy narrative arcs.[11] Martin's books appear to have been written from the beginning with this format in mind.

The novels are overflowing with episodes and characters. There are so many characters that Martin provides a list of them in an appendix at the end of each volume: in book 1, this appendix comprises eighteen pages, but by the time we get to book 5, it has grown to sixty-eight pages. Martin's acknowledgments at the end of each volume also make it clear that he is increasingly having trouble controlling the different strands of his narrative. He had originally stated that *A Song of Ice and Fire* will comprise a total of seven books, but while the second volume of book 5 was published in 2011, book 6 has not yet appeared, and Martin's fans are complaining loudly about his procrastination in various online fan sites and blogs. Television audiences, like readers, want a proper ending to the story.

Many scholars have pointed out the problems of managing complex narratives in such a way that they can be brought to a satisfactory conclusion, and some have argued that the serial format as such is inimical to closure.[12] Indeed, many television series never provide an ending, as networks simply decide not to renew them and leave the story hanging.[13] But the writers of *Game of Thrones* have promised to complete the series and bring some sense of closure to the narrative. Since Martin has not finished the books, for the last several seasons the writers of *Game of Thrones* have been relying on a plot outline supplied by him. However, even though the television series has drastically simplified the story, eliminating many secondary characters and dropping whole sequences of episodes, it still has to cover a huge multiplicity of interlacing storylines.

The multiplication of episodes that results from serialization involves the same characters in many different situations, meaning that they can easily accumulate contradictory attributes, which may lead to inconsistencies in characterization. For example, in the first book of *A Song of Ice and Fire* (as in the pilot episode of *Game of Thrones*), Sir Jaime Lannister casually throws a child out of a window to protect his incestuous love affair with his sister. Yet, by the time the narrative reaches book 5, Sir Jaime has become a chivalrous knight and a diplomatic leader of armies. If readers and viewers have not simply forgotten what the same character did in book 1, they will have to find some way of interpreting this change in order to maintain textual coherence. We can say that the character has "matured." Or we can say that the characterization is incoherent.

*Game of Thrones* has thus inherited from its source text a welter of interlacing storylines and a set of not always consistent characters. However, it still impresses viewers through the spectacular visual environment it has created.

Martin's books are full of descriptions of all kinds. A sword is "alive with moonlight, translucent, a shard of crystal so thin that it seemed almost to vanish when seen edge-on."[14] Cloth is "so smooth that it seemed to run through her fingers like water";[15] wine is "cool fire" as it trickles down your throat;[16] and unwashed men have "a sour smell."[17] There is no doubt that the fictional world of Westeros is brought vividly to life for Martin's readers.

In the television series, the depth of the reality effect carried by the written descriptions is achieved largely by careful attention to the details of setting, costumes, acting, photography, special effects, and editing that create the visual style of the show.

I do not mean that the series attempts to translate Martin's rather metaphorical style into corresponding visual images. Instead, *Game of Thrones* has created a clearly recognizable visual universe of its own, a mixture of luxurious costumes and palace interiors contrasted with dark, grubby scenes of war and violence. This visual universe has its roots in the source text, but it is much more consistent and systematic and extends over the space of the whole narrative, creating a much needed coherence and continuity, which at times conceals incongruities in characterization.

For instance, the figure below is from a scene at Castle Black, where Stannis Baratheon (Stephen Dillane) offers Jon Snow (Kit Harington) the position of Lord Commander of his army.[18] The scene shows a room with rough stone walls; an arched doorway; deeply set small, unglazed windows with wooden shutters; a fire burning in a large but plain fireplace; heavy wooden furniture; and volumes of leather-bound books. Daylight comes through the windows, and there are candles burning in standing candelabra. Three men are in the scene. One (Stannis Baratheon) is seated, with his back to the fire, in a chair padded with furs behind a heavy wooden table; the second (his follower Davos) is standing, also with his back to the fire; the third (Jon Snow) is standing in front of the table, with the doorway behind him. All are heavily dressed in black clothing. On the

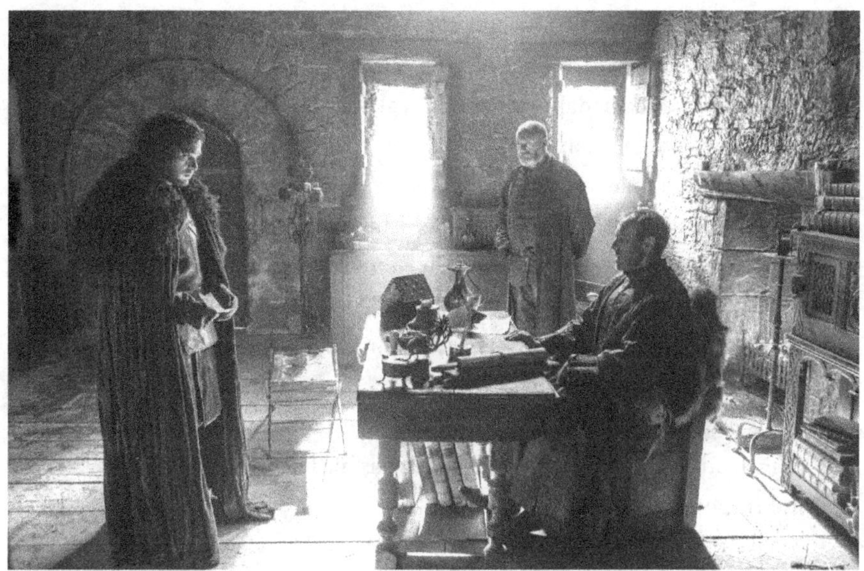

Castle Black

table and shelves are metal flasks and a metal drinking cup and documents lying open or rolled and tied. Jon holds a small paper in his hands.

What are the semantics of this image? Almost all the things we see—the stone walls, the furniture, candles, objects, clothing—indicate "oldness," something not contemporary but historical and vaguely "medieval." The stone walls also indicate a strong building, a castle perhaps. The unglazed windows, fireplace, and furnishings are functional but plain, showing a certain austerity. The fire in the fireplace and the heavy clothing of the men indicate that the season is winter and the weather is cold. Books and documents indicate organization and planning. The one man seated and the others standing could mean a hierarchical relationship; the seated man, who has documents and a metal cup in front of him, seems to be the superior of the others. The all-male environment, black clothing, and lack of color have connotations of something monastic or military, something somber, austere, and disciplined. All of these meanings are consistent with, and confirm, what we know of the plot and the characters up to this point.

We can compare this image with the figure below, from the wedding of King Joffrey (Jack Gleeson) and Margaery Tyrell (Natalie Dormer).[19] The scene is brightly lit. The back wall is covered by a tapestry with a complex woven design of heraldic figures (lion and stag). On a large table covered with two tablecloths, one white with a heavy gold fringe and the second deep red with gold embroidery and medallions, are wineglasses and metal plates with dainty pieces of something, perhaps fruit. There are two standing figures in the right side of the frame, a man and a woman. The woman (Margaery Tyrell) is wearing a long white dress with rich decorations and embroidery that leaves her back bare; her hair is elaborately arranged. The man (Joffrey Baratheon), who is quite young, is dressed in cloth of gold with black boots and belt and a mantle of gray velvet draped diagonally over his left shoulder. Both wear jewels; the man wears a crown. The woman appears to have approached the man and is looking at him, kissing his cheek while embracing him with her left arm; the man looks not at her but instead straight into the camera; he has his hand on his belt, frowns, and does not respond to her caress.

Much in this scene also signals "medievalness." However, in contrast to the previous figure, everything here—furnishings, clothing, jewels, ornaments— indicates wealth and luxury rather than simplicity and austerity. The bright light seems to imply that it is summer, and the bright colors suggest festivity and celebration. The delicate titbits of foods on the table indicate pleasure and indulgence; the dress and posture of the woman hint at sexuality. The only dis- cordant item is the stance of the king. His peculiar frown and lack of response to the woman (whom he has just married) affect our interpretation of their

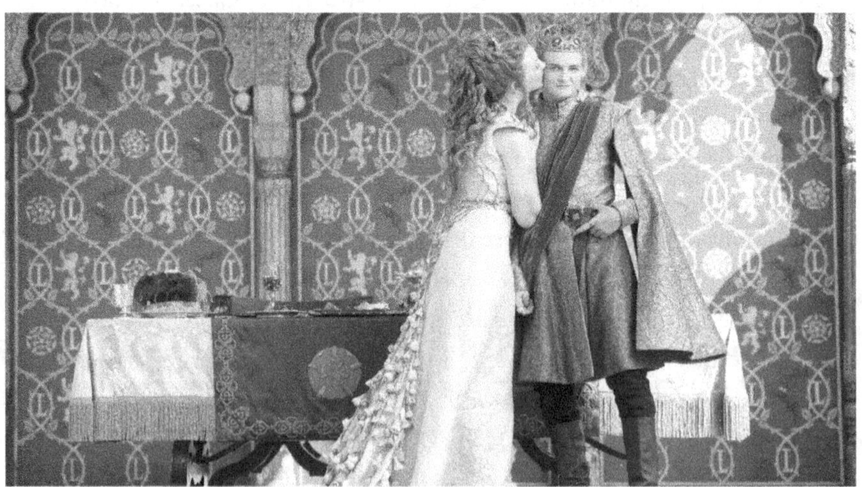

Margaery and Joffrey

relationship: her gesture acquires a connotation of supplication, while his rejection emphasizes his superiority and marks the relationship as hierarchical. Here, also, the visual environment confirms and contributes to what we know of the characters and the plot so far.

If we put the above analysis in semiotic terms, we could say that the two scenes present us with a series of visual isotopies composed of opposing terms: *cold* versus *warm, dark* versus *bright and colorful, plain* versus *luxurious, austere* versus *indulgent, abstinence* versus *sexuality.* These isotopies recur throughout the television series, with a clear geographical distribution: the north (from Castle Black down at least as far south as Harrenhal) is cold, dark, plain, austere, and generally sexless, while the south (not only King's Landing but also Pentos, the Free Cities, and Dorne) is warm, bright and colorful, luxurious, indulgent, and sexy.

This visual environment is realized with impressive consistency throughout the full eight seasons of the series. From the pilot episode,[20] Winterfell is shown as a dark place. Brown, black, and gray colors dominate; even the women are dressed in dark blue or gray. The castle is sturdy but simple, clothing warm but plain. We see little of King's Landing in the first episode except the throne room, but this has impressive architectural features: stained glass windows, a patterned floor of colored marble tiles, huge decorated pillars, and a second-story gallery of columns in orange-red marble with carved capitals. Pentos, to which we are also introduced in episode 1, is full of warm sunlight, people in brightly colored clothing, blond sandstone, large open windows, gardens, and terraces. In fact, *Game of Thrones* uses the coherence of its visual style to hold together its centrifugal narrative structure.

Visual style is essential to the viewer experience. A cinematic environment that is not faithful to the visual style that has been established for the film or television series will interfere with the viewers' participation in the fictional world. However, like action and characters, it needs to be anchored in the structure of the narrative. In Jackson's *Lord of the Rings*, the visual appearance of Rivendell gives us significant information about the elves and their role in the story. In *Game of Thrones*, on the other hand, the visual environment, however consistent, does not seem to play any part in the thematic structure. The fact that Winterfell is cold, bleak, and dark does not correspond to any clear function of Winterfell in the narrative.

There are some isotopies that all places in Westeros have in common, and these often serve as indicators of important themes. One is, in fact, the isotopy of "medievalness." Another is a *spatial* isotopy of *indoors* versus *outdoors*, in which indoors is divided into *private* versus *public spaces*, and outdoors into *cities and gardens* versus *wilderness* (frozen wasteland, forest, steppe, or desert). The division into private versus public spaces is related to the theme of *deception* (isotopy of *deceit* versus *loyalty*) in the narrative: much of the story is constructed around an endless series of secret plots and counterplots. But more significant than these is a virtually universal isotopy of *hierarchy* (*superior* versus *inferior*), related to the theme of *power*. Between men, hierarchy takes the form of feudal or military relationships. Between men and women, it frequently takes the form of sex (sex, in *Game of Thrones*, is a form of the exercise of power, unless you make the mistake of falling in love, in which case you simply become a victim). The exercise of power implies physical violence, which can occur at any time and in any place in the narrative. Power is probably the central theme of the whole series (as indeed its title indicates), to which even family relationships seem to be subordinated: power is rooted in the position of the family, and the loyalty of family members is its surest support. Loyalty, however, is a dubious foundation, since most of the characters in the series are ready to betray one another at a moment's notice.

*Game of Thrones*, I would argue, is an example of an adaptation that succeeds in achieving a more coherent narrative than its source, through ruthless simplification of the plot and a consistent visual environment; interestingly, this is due to the show being "unfaithful" to its source. Its consistent visual style confirms and reinforces what we know of the characters and the development of the plot, and it is impressively spectacular. But in itself, it is not enough to give semantic coherence to its centrifugal story. We cannot say that the bleak and austere environment of the north is meant to convey some particular quality, for example, moral superiority over the luxurious and indulgent south (in fact, some of the most appalling violence in the whole series takes place in the north). The result is that the visual world of *Game of Thrones*, like the verbally created world of *A Song of Ice and Fire*, leaves us mainly with an impression of senseless and chaotic violence.

The producers of *Game of Thrones* have worked hard to develop it into a transmedia storytelling franchise, mostly by encouraging speculation among fans about how the story will develop. Instead of basing their own decisions on what the fans want, however, they seem to have deliberately chosen to create the most startling plot twists they can devise; character consistency and plot coherence are secondary. What holds the series together, along with the spectacularity of the visual style, are these sudden reversals of viewers' expectations. But the coherence thus achieved is superficial. It does not help us to bring the strands of the scattered storylines together into a satisfactory conclusion.

## Narrative Structure and Visual Semantics in Space: *Star Trek*

If we want to examine how narrative structure and the semantics of the visual setting interact to construct coherence, a good example would be *Star Trek: The Motion Picture*.[21] The team behind the film had already shared years of experience and cooperation in the production of dozens of episodes for the original television series, and they had an established fan base of informed and demanding viewers. The desire to maintain continuity with the television show was so strong that the production team went to considerable lengths to assemble the original actors to play the same characters. So we have reason to expect both strong continuity with the television series and a coherent narrative.

The overarching storyline of the film is based on the mission of the starship Enterprise: to protect Earth by confronting and hopefully destroying V'GER, an all-powerful, alien entity. Within this storyline, the main characters each have their own "narrative program" (in Greimas's terminology), and the main plot is a tightly interwoven network of these individual narrative programs. Spock's (Leonard Nimoy) desire for contact provides most of the information on what V'GER is (and, incidentally, an answer to Spock's own quest). Kirk's (William Shatner) actions as captain demonstrate the implied claim that experience, cooperation, and an open mind are more important than the latest military technology when dealing with the unknown (though the need for up-to-date knowledge of technology is also recognized). In fact, it is the collaboration of all the main characters that provides the successful resolution of the plot. The narrative structure of *Star Trek* is thus strongly coherent.

The semantics of the narrative, developed in its visual environment as well as in dialogue and action, are also coherent. The very first sequence of the film, showing three Klingon spaceships attacking a force field, establishes a powerful isotopy of "spaceness," or better yet, "science-fictionality." This isotopy is lovingly developed throughout the film, with nostalgic visual references to the old television show (isotopy of "*Star Trek*-ness") and extended shots of the Space Station, the Enterprise being refitted, transport modules coming and going, views of Earth from space, and long graphics sequences of the hostile force field. But it also has a functional role in the story. It divides the environment of the film into *space* versus *Earth*. Space is shared by the Galactic Federation, made up of humans and their humanoid allies, and the Klingons, their traditional enemy. The Klingon spaceships have an all-male, exclusively Klingon crew, and their standard modus operandi is aggression; on the other hand, Federation installations consistently have sexually, racially, and ethnically mixed crews whose preferred method is friendly contact and communication. We could diagram this environment as in the following table:

### *Diagram of the isotopy of "spaceness"*

space

/ \

| Klingon | vs. | Federation (humans and humanoid aliens) |
|---|---|---|
| all-male, Klingon only | vs. | sexually, racially, and ethnically mixed |
| aggression | vs. | communication |
| isolation | vs. | cooperation |

The main narrative program of the film is organized around the opposition *machine* versus *human*. This opposition is introduced almost casually, and its importance does not become clear until the identity of V'GER is established, something that depends on a gradual, piecemeal revelation and interpretation of information. All the main characters of the film participate in this data-gathering process; cooperation is an essential quality of the humans and humanoids of the story, who function as a group, while V'GER is alone. V'GER is aggressive, threatening to exterminate the "carbon units"; humans respond by trying to understand and communicate. V'GER has questions, is curious, searching; the humans have the answers it needs. V'GER is in its infancy; humanity is mature. V'GER has logic; humans have the ability to go beyond logic. V'GER has knowledge; humans have purpose. The story is built on a systematic set of semantic oppositions, as seen in the following table:

| Machine (V'GER) | vs. | human/humanoid ("carbon units") |
|---|---|---|
| ↓ | | ↓ |
| isolation | vs. | cooperation, friendship, loyalty |
| aggression | vs. | communication |
| infancy | vs. | maturity |
| questions | vs. | answers |
| logic | vs. | intuition, leap beyond logic |
| incomplete | vs. | creating their own purpose |

The opposition between conscious machines and humans/humanoids is mediated by the Vulcan, Spock, who understands the thought processes of the machine. The solution, the union that allows the machine to fulfill its mission, is mediated by the Deltan, Ilia (Persis Khambatta). Both are aliens with extra-human abilities and allies in the Galactic Federation. The implication of the final conversation between Kirk, Spock, and Bones (DeForest Kelley) on board the Enterprise seems to be that the Galactic Federation will recognize conscious machines as another life form and will seek alliance, or at least peaceful coexistence, with them.

As we see, the story centers on the isotopies of *aggression* versus *communication* and *isolation* versus *cooperation* that were introduced in the very first visual sequences. Both the Klingons and V'GER propose to solve their problems by isolation and aggression. In contrast, all the efforts of the Enterprise and its crew are collaborative and directed at how to establish communication; how to find a common language; and how to understand the other, respond to it, and ultimately accept it. This process, it is implied, will allow humans, humanoid aliens, and conscious machines to create their own purpose, making their existence meaningful.

## Stories and Storyworlds

It is clear that *Star Trek: The Motion Picture* is a very coherent text. The same semantic structures that establish the visual environment also govern the narrative programs of the characters. The same isotopies and themes recur in visual storyworld, action, and dialogue. The resolution of the plot corresponds to the problems posed in the initial visual sequences, not only in the sense that the original threat is averted but also in that this is achieved through the qualities—such as cooperation, communication, and multiethnicity—that we saw in the Federation allies in the initial scenes. The film can be enjoyed with no prior knowledge of its source text, but it chooses to maintain close continuity with it, making nostalgic visual references to the television series.

In fact, one wonders if there isn't rather too much continuity. The film plot is very close to a frequently used plot formula of the old television show: humans encounter alien life forms that at first appear threatening, but due to the combined skills of the Enterprise crew, conflict is replaced by communication and understanding. The film updates this recipe by introducing a conflict with a life form that is a machine, leading to a reconfiguration of the semantic themes used in the television show (though there had been episodes involving cyborgs and computers, they had generally been presented as threats to be destroyed), but for those spectators who are not passionate fans, and in spite of the digital graphics of the 2001 director's edition, the result feels a bit too familiar.

Too much continuity can become a problem in transmedia franchises that rely on repetitive narrative structures, such as superhero comic book adaptations. Sequels, prequels, and spin-offs tend to use the same ingredients, unless they have been planned from the beginning as parts of a unified narrative sequence. George Lucas's original *Star Wars* trilogy (1977, 1980, 1983) was planned as a whole, and the original story was then expanded in prequels and sequels, using the same visual storyworld and many of the same actors. However, instead of simply repeating the tale of daring rebels fighting technologically superior militaristic enemies, the later films introduce new narrative twists. They maintain continuity with the original story but without repeating it in each new film (although the later spin-offs are much more reliant on a formulaic plot). The *Harry Potter* films (and the books they are based on) are another example of an original story becoming more complex as the sequence of films develops.[22] A franchise cannot rely exclusively on a familiar storyworld, however attractive. New additions need to have something new to say—a new story—to maintain engagement with the audience.

In conclusion, this chapter's discussion has shown how the concept of semantic isotopy can be used to analyze visuality. Visual style is usually defined in terms of the external factors that go into producing it. I believe it can also be analyzed in terms of its internal semantic constituents—its isotopies—and their interpretation by the audience.

Finally, I would like to return briefly to Ryan's distinction between storyworld and plot, the spatial and the temporal components of a narrative, quoted in the beginning of this chapter. Ryan is right, of course, that any narrative needs to have both. However, I am not so sure that we can or should divide stories into world centered and plot centered. The point I have been trying to make is that these two aspects are—or should be—interconnected. In a successful film or television narrative, the visual environment is not just the space in which the plot develops but it is also structurally linked to the same semantic elements— the same isotopies—that operate in the plot, and for this reason it has a very significant contribution to make to the consistency and coherence of the story.

Ryan also suggests that the popularity of science fiction and fantasy narratives in transmedia storytelling has to do with the fantastic, nonrealistic nature of their storyworlds. "The more a storyworld departs from the real world, the greater the cognitive effort needed to imagine it. This may explain why fantastic and science-fiction worlds tend to generate more transmedia activity than realistic everyday worlds."[23] I disagree. I enjoy a story set in a science fiction or fantasy world, but I would be quite happy to go on to enjoy an entirely different storyworld, if it is equally visually spectacular, coherent, and consistent. The impulse to prolong the narrative experience originates not in the cognitive effort needed to understand its visual world but in the feeling that this particular story has not exhausted the potential of that world, that it must have more stories to tell. This, it seems to me, is not unlike the impulse involved in any adaptation: this text, this storyworld, has more stories to tell.

## Notes


1. Henry Jenkins, *Convergence Culture: Where Old and New Media Collide* (New York: New York University Press, 2006), 95.

2. Ibid., 114.

3. Marie-Laure Ryan, "Transmedia Storytelling as Narrative Practice," in *The Oxford Handbook of Adaptation Studies*, ed. Thomas Leitch (New York: Oxford University Press, 2017), 538–39.

4. Algirdas J. Greimas and Joseph Courtes, *Sémiotique: Dictionnaire raisonné de la théorie du langage,* volume 1 (Paris: Hachette, 1979) and volume 2 (Paris: Hachette, 1986).

5. Algirdas J. Greimas, *Sémantique structurale* (Paris: Larousse, 1966). In the 1990s and early 2000s, there was considerable interest in the possibility of using the narrative theories of semioticians, such as Roland Barthes, Claude Brémond, and Gérard Genette, in adaptation studies. Though this interest has left traces in the vocabulary of the field (see Thomas Leitch, "Twelve Fallacies in Contemporary Adaptation Theory," *Criticism* 45, no. 2 [2003]), the narratology of Greimas has not received the attention that I feel it deserves; a notable exception is Betty Kaklamanidou, *The "Disguised" Political Film in Contemporary Hollywood: A Genre's Construction* (New York: Bloomsbury, 2016).

6. As Wolfgang Iser has pointed out, making sense is an active process: the audience is never simply a passive receiver of ready-made meaning but has to draw on their full semiotic competence, their knowledge of the codes and conventions of the medium, as well as the cultural values and assumptions behind the text; in Linda Hutcheon *A Theory of Adaptation,* 2nd ed. (Oxon: Routledge, 2013), 134.

7. This applies to fictional worlds that are reasonably close to our own culture and society. To interpret a fictional world created by a different society, we need to be familiar with the codes and conventions of that culture.

8. See Bradley J. Birzer, *J.R.R. Tolkien's Sanctifying Myth: Understanding Middle-earth* (Wilmington, DE: ISI Books, 2003); and Jeff Peterson, "Why Peter Jackson's *Lord of the Rings* Succeeded as an Adaptation," *Deseret News*, December 7, 2012. https://www.deseretnews.com/article/865568286/Why-Peter-Jacksons-Lord-of-the-Rings-succeeded-as-an-adaptation.html.

9. Thomas Leitch, *Film Adaptation and Its Discontents* (Baltimore: Johns Hopkins University Press, 2007), 127–50.

10. Jason Mittell, *Genre and Television: From Cop Shows to Cartoons in American Culture* (New York: Routledge, 2004), 25.

11. Virtually all discussions of quality television drama series agree on the central importance of high production values and complex story lines; see Janet McCabe and Kim Akass, *Quality TV: Contemporary American Television and Beyond* (London: I. B. Tauris, 2007); Gary R. Edgerton and Brian G. Rose, *Thinking Outside the Box: A Contemporary Television Genre Reader* (Lexington: University Press of Kentucky, 2005); Ethan Thompson and Jason Mittell, *How to Watch Television* (New York: New York University Press, 2013); and Elliott Logan, *Breaking Bad and Dignity: Unity and Fragmentation in the Serial Television Drama* (New York:

Palgrave Macmillan, 2016). The expression "high production values" is in some ways unfortunate, because it seems to imply that quality depends simply on the amount of money spent. Of course, money is important: it is necessary to ensure good screenwriters, directors, actors, photographers, and special effects designers, and it can pay for sumptuous costumes, impressive settings, and filming on-site in exotic locations. However, the size of the budget as such does not guarantee the results. For the term "signifiers of quality," see Sarah Cardwell, "Is Quality Television Any Good?" in McCabe and Akass, *Quality TV*, 29; and Roberta Pearson, "*Lost* in Transition: From Post-Network to Post-Television," in McCabe and Akass, *Quality TV*, 255.

12. See Ellen Seiter and Mary Jeanne Wilson, "Soap Opera Survival Tactics," in Edgerton and Rose, *Thinking Outside the Box*; and Sean O'Sullivan, "*The Sopranos*: Episodic Storytelling," in Thompson and Mittell, *How to Watch Television*. See also the discussion in Logan, *"Breaking Bad" and Dignity*.

13. Henry Newcomb, "Reflections on *TV: The Most Popular Art*," in Edgerton and Rose, *Thinking Outside the Box*, 31.

14. George R. R. Martin, *A Game of Thrones* (London: Harper Voyager), 8.

15. Ibid., 25.

16. Ibid., 120.

17. Ibid., 114.

18. *Game of Thrones*, season 5 (2015), episode 2, "The House of Black and White."

19. *Game of Thrones*, season 4 (2014), episode 2, "The Lion and the Rose."

20. The important semantic codes of a text tend to appear in the beginning; see Roland Barthes, *S/Z* (Paris: Seuil, 1970), 23–27.

21. Originally released in 1979; director's cut with new graphics released in 2001.

22. The repetitive plots of the superhero franchises have an appeal of their own. Umberto Eco (*The Role of the Reader: Explorations in the Semiotics of Texts* [Bloomington: Indiana University Press, 1979], 17–120) has pointed out that a repetitive plot structure is not necessarily a drawback: we read detective stories by our favorite authors (or watch the latest BBC *Sherlock Holmes* adaptation), because we are looking for the same plot patterns, the same characters, and the same written or visual style that we have enjoyed before.

23. Ryan, "Transmedia Storytelling," 539.

III

# ElasTEXTity and Adaptation

7

# Transnational Adaptations in a Globalized Context

## The Case of the Greek Film Musical

Ursula-Helen Kassaveti

IN THE TWENTY-FIRST CENTURY's film industry, it all comes down to transnationality, both in structural and stylistic elements. Far from being a mere aphorism, such a statement sums up the results of the never-ending transformations in modern adaptation practices. After World War II, various non-American films, from the Turkish *3 Dev Adam / Captain America & Santo vs. Spider-Man* (1973) and the Greek *I Love You / S' agapo* (1971)—an adaptation of *Love Story* (1970)—to the Italian *Alien 2: Sulla Terra* (1980), emerged as transnational adaptations of their archetypical American counterparts, functioning as respectable domestic film products targeted toward a specific national audience. Ignoring issues of fidelity to their Hollywood origin, as well as their respective literary source, such films operate on a multifaceted and contested plane, where globalization, cultural appropriation, and the reception of a reconstructed "dominant" discourse can play a crucial role in shaping and promoting a local film industry.

Such globalized transnational "mediascapes," to use the term suggested by cultural anthropologist Arjun Appadurai in his seminal book *Modernity at*

*Large,*[1] defined as various forms of media and their images distributed world-wide, provide a framework for examining and analyzing localized filmic versions in a different light. As Appadurai argues, homogenization or Americanization could be perceived as forces that resist globalization, to the extent that the new cultural economy, with all its discontinuities and disjunctures, cannot be perceived as linear and simply ordered, especially in the international media sector, where various transnational multimedia structures meet.[2]

A further set of potential consequences stemming from the aforementioned advancements can be traced within the geographies of such transnational "mediascapes": what is chosen to be adapted in country X and for which audience. It would be an oversimplification, and almost a cliché, to argue that transnational adaptations are a common practice within the film industry of Eastern Asia[3] or Europe. In the context of transnational adaptations, this chapter aims to propose new outlines and pose fresh questions about the characteristics and limits of this practice, based on the paradigm of a small national European cinema: the Greek one.

Drawing upon Appadurai's theory of globalization and the concept of "mediascape,"[4] Anthony Pym's examination and understanding of globalization versus localization in the moving text,[5] and, particularly, Iain Robert Smith's interpretative model of transnational adaptation,[6] I use the various transnational articulations of the Greek film musical genre in the 1980s as my case study. The 1980s Greek film musical, as well as the whole genre in general, constitutes an exploratory, rare, and, subsequently, problematic terrain in regard to transnationality as it stands in contrast to other Greek popular genres that usually adapt Greek sources. The Greek film musical bears an American influence in terms of form (mainly choreography and iconography), which constitutes instances of transnational adaptation in this popular genre. However infrequent these transnational moments are, they do occur and therefore merit examination. By focusing on different aspects of transnational adaptation (i.e., agreement and deviation from the original, moments of subversion, transnational yet localized and vernacular character) in this case study, my ultimate goal is to provide insight into the cultural implications deriving from this practice and to propose a model of examination deploying the 1980s Greek film musical as a case study. In this way, I aspire to provide motives for further critical and academic exploration of popular transnational cinematic texts, or rather "mediascapes."

# Mediascapes among a Contested Terrain

The term "transnational" is best understood alongside different but interrelated concepts, such as Americanization, globalization, and localization. I offer here a brief commentary based on the growing literature,[7] because I find it important to understand why these terms are always in negotiation due to the ever-changing nature of global economy and politics. The Frankfurt School's[8] demonization of popular culture's Americanization (primarily in the Western world) and its supposed by-products serving as a remedy for "dumb" consumers, or the conspiracy theory of cultural imperialism and the ramifications of such interpretations, have long been confuted by cultural theorists, such as sociologist John Thompson. Thompson asserts that this argumentation is both outdated and unconvincing, as "the reception and appropriation of cultural phenomena are fundamentally hermeneutical processes in which individuals draw on the material and symbolic resources available to them."[9] Moreover, he insists on the value of the interpretative process, which helps people to make meaning of the received messages as well as their everyday lives. In short, individuals tend to produce their own ephemeral meanings through constant interaction with various cultural products, which they in turn use to make sense of their everyday world, and are not necessarily guided or alienated by them according to their dominant ideological models. This also implies that Americanization cannot be perceived separately from globalization and localization.

The emergence of multiple "mediascapes" within a global and a local context can be regarded as a direct consequence of the fluid cultural terrain of the two previous centuries. These "mediascapes"

> refer both to the distribution of the electronic capabilities to produce and disseminate information (newspapers, magazines, television stations, and film-production studios), which are now available to a growing number of private and public interests throughout the world, and to the images of the world created by these media. These images involve many complicated inflections, depending on their mode (documentary or entertainment), their hardware (electronic or pre-electronic), their audiences (local, national, or, transnational), and the interests of those who own and control them.[10]

It is in this context of distribution, of public and private interests and of disparate cultural flows, that these "mediascapes" come to life and spread "out-of-scale transnational images that are now reorienting international social discourses and processes."[11] So, it is difficult to deny that dominant hubs and peripheral outlier regions[12] in cultural production usually coexist in harmony. In view of this, Pym suggests that, while exploring issues of translation and distribution, globalization can be seen as "one wide process," of which "internationalization" and "localization" are its main parts. He further argues that "in order to global-ize, you first make your product general in some way ('internationalization'), then you adapt ('localize') to specific target markets ('locales')."[13] Americani-zation should not therefore be considered as the only generative cultural force of "mediascapes" but rather as a parallel, intersecting, and decisive factor in contributing to local markets. This view clearly represents a departure from an understanding of global cultural economy "in terms of existing center-periphery models"[14] to a view of the first as an elaborate and overspreading order of disjunctures.

## Models of Transnational Adaptation

The special character of transnational adaptations as "mediascapes" and agents of global cultural economy may be traced both to television and film. In the first case, Michele Hilmes distinguishes four types of transnational produc-tions, which, apparently, adhere to the transnational demands of a vast and multifocal market and relate to the audience's needs for a localized version of an international production.[15] Regarding film, Darrell William Davis discerns three distinct types of "national cinema studies."[16] He labels the first type "reflectionist," as it "reflects conditions in the culture that produced the films."[17] What precisely constitutes a national cinema as "reflectionist" stems from its national specificity and its typical, or even stereotypical, filmic representations of national psychology. The second type is the "dialogic" national cinema, which underlines the explicit dialogue that exists between each national cinema with other cinemas. According to Davis, the "dialogic orientation is based on a textualist assumption rather than on a culturalist one"[18] and connects different film industries. The third type, "contamination," perceives national cinema as a field where syncretism and productive processes take place. Davis does not view "contamination" as a negative process. On the contrary, he argues that such an approach offers an enlightening explanation about the constant ties and correlations between different national cinema cultures. As Davis further suggests, since contamination models are not constructed on binary oppositions (i.e., black versus white, east versus west), a national cinema cannot be solely considered as a one-way reflection of a culture or a dialectical and intertextual relation between cinemas and cultures,[19] but a cinema based on reflections and dialogues, which further promote its development.

Finally, Iain Robert Smith proposes a "comparative model of transnational adaptation,"[20] which is based on the two following basic traits of transnational adaptations:

> They are based on the "transnational" character of Hollywood films.
> They are prone to be faithful to a particular film text and not to a literary source, and they can be explored and highlighted through a case study examination.

Smith claims that transnational adaptations should be examined under specific perspectives: the reasons an adaptation "proliferates" from a particular archetype; the ways governments and industries influence the adaptations made, especially in the case of copyright laws and so forth; and, finally, the meanings of the latter within the framework of production and reception. It is this model that I further employ and apply in this chapter. On the basis of the triptych proliferation/industry/reception, I explore similar issues in the popular Greek film production of the 1980s. The investigation of a local film market can provide us with an understanding of the way the popular Greek film industry operated and how it approached the international market, as well as illustrate the need for a localized adaptation that would share the merits and the allure of a renowned film production and provide them with specific cultural insight. Greek popular cinema, primarily the Greek film musical genre from the 1960s to the 1980s, and its transnational adaptation strategies constitute our general point of reference, while our focus is the Greek transnational adaptation of *West Side Story* (1961), titled *Otan oi Rodes Horevoun / When the Wheels Are Dancing* (1984).

# Old Greek Cinema and the Transnational

In general, Greek cinema is divided between the Old Greek Cinema (Palaios Ellinikos Kinimatografos, or PEK) and the New Greek Cinema (Neos Ellinikos Kinimatografos, or NEK). Overlooking some minor differentiations regarding periodization, Paradeisi[21] and Delveroudi[22] note that from its early stages in the 1950s, PEK would reach its highest popularity and high production numbers during the 1960s, facing a temporary decline in the 1970s due to the advent of television. At the end of this decade, a Greek popular production revival led to the box office successes of the 1980s, up until the emergence of the local direct-to-video circuit[23] that actually marked its decline and accelerated PEK's preannounced yet sluggish death.

PEK consists of a variety of genre films (comedy, melodrama, musical, etc.) produced by Greek production companies and private entrepreneurs. Such films were mainly based on a quick production pattern and a local star system, with an (usually but not always) intuitive syntax, commonplace subject matters, and a Manichean structure, which usually led to a "happy ending" and kept the audience entertained. The Greek film musical was one of the most popular film genres of that time. Although it couldn't compete with the American genre or finance any relatively expensive film productions, Greek producers and filmmakers sought to transcribe it in Greece, either as a poor relative of the American one or as a fresh and creative force to open up a new road for popular entertainment.[24] Moreover, it portrayed different tensions emerging in the Greek society, for example, the tension between conservatism and modernity.[25]

It should be noted that Old Greek Cinema was born as the country started to leave World War II, a civil war (1946–49), and their consequences behind. During that time, "the tertiary sector expanded rapidly, the feeble manufacturing sector came almost to a standstill, and agriculture remained badly organized."[26] During a climate of political instability that culminated with the 1967 coup d'état, the Greek film industry was not a priority for the state. The official overall approach toward the domestic film production was indifferent, while its legal frame was fragmentary, deficient, and bureaucratic—as opposed to many European countries whose laws reinforced their national film production.[27] In addition, high taxation and censorship practices have proved to become long-term enemies of the Greek film industry.[28] Greek theaters were inundated with foreign films, mainly American, thanks to the establishment of domestic distribution companies after 1948. Due to currency flow that increased steadily and the special conditions of the American film distribution[29] in Greece, the number of American films in the Greek market rose throughout the 1960s (79 films during 1926–28 versus 326 during 1963–64).[30] Although attempts were made toward a more systematic and richer production of Greek popular films, American production films dominated the market, as their success was considered foolproof.

Greek films were influenced by both European and American productions. Regarding adaptations, first, we encounter commercially successful adaptations of Greek popular plays,[31] as well as loose adaptations of European plays (e.g., *Nychta Gamou / Wedding Night* [1967], based on Georges Feydau's *L' Hôtel du libre échange*) and even fairy tales (e.g., *I Chionati kai ta Epta Gerontopalikara / Snow White and the Seven Spinsters*, a modernized adaptation of the Brothers Grimm fairy tale). Second, one can trace indirect transnational influence from Asian (melodrama) and American genre films (mainly musicals). Greek melodramas often deploy the form of *Yeşilçam*[32] and similar Indian films.[33] The American "juvenile delinquency" cycle from the early 1960s was also appropriated in Greece. Director Giannis Dalianidis's films (e.g., *Katiforos / The Decline* [1961] and *Nomos 4000 / Law 4000* [1962]), especially, not only echo the impact of such films as *The Blackboard Jungle* (1955) but also comment on the imaginary contradictory nature of Greek society and expose the filmmaker's moral panic against the supposedly ill-natured Greek teens of the decade. Dalianidis also shot a series of film musicals in the 1960s that, notwithstanding their popularity, underscore an influence from the respective Hollywood tradition in terms of numbers or set design (although made on a much lower budget). Third, there are transnational adaptations, more or less faithful to a preexistent film prototype, that localize the content: popular Greek film either explicitly or implicitly used or humorously reconstructed American or European films (e.g., *S' agapo / I Love You* [1971] is a transnational adaptation of *Love Story*, and *Hameni Eytyhia / Lost Happiness* [1966] is an adaptation of *Les Parapluies de Cherbourg* [1964]). Finally, there are transnational adaptations in the form of parody, especially in the Greek direct-to-VHS circuit, such as *Alles to Protimoun Gouli / Others Like It Bald* (1986), which is a free adaptation of *Some Like It Hot* (1959).

# Otan oi Rodes Horevoun / When the Wheels Are Dancing (1984): A Case Study

The reasons behind the choice of the Greek film musical of the 1980s as a basis for the examination of transnational adaptations are the following: First, it belongs to a highly popular genre and is therefore a suitable terrain to examine how well-known films are transnationally adapted in local, vernacular versions, without losing touch with the original's archetypical structure. Second, singing and, mainly, dancing as cultural phenomena are staples of Greek culture. Despite the many turbulences Greek society has endured after World War II and during the 1960s,

> dance remains a central component of many celebrations. . . . The cultural emphasis throughout Greece on bodily presentation . . . lend[s] a special interest to the ways in which Greek men and women present themselves and evaluate each other's actions in the contexts of dance.[34]

Third, its release marks the return of Dalianidis to the film genre by which he was established after almost a decade. Of course, this return also coincides with the rebirth of popular cinema, as I will shortly argue, and the rise of new subject matter, film cycles, and even social conditions. Within this framework, it is interesting to find out whether the classic Greek film musical of the 1960s can adhere to a specific sociocultural snapshot or must allege itself as an outdated film genre.

Finally, the beginning of the 1980s witnessed a brief renaissance of Greek popular cinema. Dalianidis made a comeback, shooting his most famous social drama trilogy—*Ta Tsakalia / The Jackals* (1981), *I Strofi / The Turn* (1982), and *Oi Epikindynoi / The Dangerous* (1982)—and launching what I refer to as "the film cycle of delinquency."[35] The filmmaker further developed youth themes in other popular genres, thus marking the return of the musical. This time, however, the musical reappeared as a transnational adaptation of an older, successful 1957 Broadway stage production by Arthur Laurents and then a film by Robert Wise and Jerome Robbins (1961): *West Side Story*.

After the box office success of his first film musical in the 1980s, *Kamikaze, agape mou / Kamikaze, My Love* (1983), Dalianidis shot *Otan oi Rodes*

*Horevoun* / *When the Wheels Are Dancing* (1984). *Rodes* constitutes not only a transnational adaptation of *West Side Story* but also an adaptation of a film adaptation that was adapted from a Broadway musical, which in its turn is an adaptation of Shakespeare's *Romeo and Juliet*.

Dalianidis resorts again to juvenile delinquency and its consequences upon the love of two different characters: the religious and fragile beauty Eirini (Vassia Panagopoulou) and Giorgos (Stamati Gardelis), a handsome and rebellious punk who lives in a squat with his gang, wasting his time on motorbikes, feuds with rival gangs, his band, and attempts to create the loudest possible disturbance in his neighborhood. Built on a dual-focus as well as a multi-focus narrative, the film, as a modernized version of the 1961 American film and Dalianidis's successful musicals from the 1960s, revolves around ten rhythmic numbers. However, only six of them adhere to the typical singing/dancing routine. This is due to a number of reasons: first, some *West Side Story* numbers couldn't find a place in the Greek version (e.g., the "America" number would be irrelevant to the *Rodes* plot); second, other numbers, such as "Cool," needed elaborate orchestration and choreography, both of which couldn't be delivered due to the popular Greek cinema's quick and rough production routines as well as budgetary restraints.

*West Side Story*'s basic plot, typecasting, and conventions, as well as some of its characteristic dancing and singing sequences, are either loosely or more strictly adapted "à la grecque" by the Greek filmmaker: a forbidden love is born between a prominent gang member and the sister of a typical Greek young man, whose fellow Christian Orthodox parish members surprisingly operate as a gang. Instead of representing two different ethnic stereotypes, Dalianidis bases the conflict on an antithesis between two particular age groups and cultural opposites detected in Greece:[36] the younger "punks" and the older Christian Orthodox (and some of their sympathizers, like Eirini). Through various conflicts and episodes culminating in Giorgos's attempt to kill Eirini's brother (Pavlos Evangelopoulos), the two lovestruck protagonists end up together and their friends reconcile.

Thanassis Bikos composed rock-style music, while the film also features prerecorded tracks. Dalianidis's establishing sequence with Giorgos singing "Ego Den Thelo Merokamato" / "I Don't Want to Work" resembles *West Side Story*'s "Something's Coming" sequence. Yet, the *Rodes* musical number, which establishes the film's characters and their offensive behavior, is only "choreographed" in terms of editing and does not include actors actually dancing, alternating between multiple close-up shots of Giorgos singing and long shots of him and his gang on their motorbikes. On the contrary, the original's "Gee, Officer Krupke" sequence appears almost identical in *Rodes*: in a multi-focus narrative, all the gang's members apologize to a Greek police officer about their self-described "problematic behavior" ("Eimaste Paidia Provlimatika" / "We Are Problem Children"). Eirini and Giorgos sing "Ti na sou Krypso, Agape mou" / "What Can I Hide from You, My Love" in a subsequent paired-scene arrangement, resembling "Somewhere" sung by Tony and Maria. The song "I Agape mas na zei" / "Our Love Should Live," sung by Giorgos underneath Eirini's balcony, resembles the "Tonight" number from *West Side Story*, which, in turn, refers to the Romeo and Juliet balcony scene.

The "traditional" notion of choreography and dance included in the American genre is changed significantly. The film's basic characters limit themselves to simplistic dance moves—they aren't professional dancers—while the secondary characters, for example, Giorgos's gang, dance according to a more elaborate choreography, showcasing their skills. As no archival material could be located (other than the film's promotional posters and ads), both singing and dancing sequences could be a hint of the filmmaker's passion for adapting a classic Hollywood film musical, but apparently he had to make do with what came in handy: for example, he chose actors and actresses with average singing and dancing abilities (meaning also that perhaps they hadn't undergone special music or dance training for the film, apart from the specific scenes during which they engaged in musical or dance acts).

In addition, *West Side Story*'s plot undergoes substantial changes in *Rodes*: the first encounter of the couple, characterized by an almost dystopian and hyperromantic haze, does not take place in a gym ballroom dance scene but during a multi-focus number, where the Christian Orthodox members of the neighborhood church accuse the gang of disturbance and attack them, singing "I Ellada pote den Pethainei" / "Greece Never Dies." The *West Side Story* ballroom scene is replaced by a breakdance competition in a disco, where rival gangs, like the "Kalamatianos" gang (which does not interfere with Giorgos's relationship or his gang), dance alongside professional dancers. This choice echoes the influence of contemporary musicals and MTV musicals,[37] such as *Flashdance* (1983) and *Breakin'* (1984)—the latter lending its song "There's No Stopping Us," by Ollie & Jerry, to *Rodes* for a break-dancing number. Due to the

Greek film musical's conventional happy ending, *Rodes* does not end with the death of Eirini's brother. Instead, the filmmaker reserves a grand reconciliatory finale for the audience, which, according to Papadimitriou, is "a staple of the Greek musical with Dalianidis' films."[38]

Both protagonists become the embodiment of two antithetical ideological values that still pervade Greek society: tradition, as expressed via Orthodoxy and faith, and modernity, as conveyed by the rebellious and unstoppable youth of the decade (even if it shows signs of deviance). However, in the end, opposing sides seem to reconcile, especially after Petros's injury by Giorgos. The Greek film musical is characterized by two opposite identities: an "introverted" one that explores Greekness and the tension between the conservative and modern and an "extroverted" one that operates within the context of postmodernity. *Rodes* appears to be introverted in its content, reflecting Dalianidis's persisting views on youth and rebellion, while it appears extroverted to the extent that it adopts a postmodern form, bearing influences from both *West Side Story* and popular film musicals of the 1980s and their main characteristics (clothing, music, video-clip influence). It adopts the classical Greek film musical's structure, which in turn follows the respective American archetype, adapts it to the 1980s cultural and cinematic instance—for example, the outbreak of juvenile delinquency or the popularity of motorbikes—and, finally, transforms it into a new vernacular, yet transnational, mediascape.

## Toward a Transnational "Localization"?

In *Otan oi Rodes Horevoun*, Dalianidis exploits a gamut of devices in order to adapt and further modernize the paradigmatic *West Side Story*. Despite the differences, this transnational adaptation provides insight into the ways a local film market utilizes a foreign source to make it vernacular again. *Rodes* fits into Smith's triptych of proliferation/industry/reception, as it is based on a transnational text, which has been heavily adapted, and also verifies the filmmaker's interaction with the latter and its incorporation into a modern context. What is also important is that nobody involved in the Greek production conformed to any copyright laws concerning the film or the stage play. It was under a "silent agreement" that *West Side Story* was adapted and localized, and nobody, except those who had prior knowledge of *Romeo and Juliet* or *West Side Story* (the film), could identify *Rodes*'s primary source. Finally, as far as reception is concerned, *Rodes* reached twelfth place among thirty-eight films with 77,140 tickets during its release in the 1984–85 season. Taking into account that the popular Greek cinema's downward spiral had already started and that *Rodes* was the only musical of the season, it could be considered as a moderate success, which allowed Dalianidis to use the protagonist couple in several direct-to-VHS films and TV series that followed.

Also of interest is that a transnational adaptation such as *Rodes* doesn't have to strictly follow the production schedule of the Hollywood film industry: a transnational version of an American or other box office success could appear either around the time of the release of the original or asynchronously, as was the case with *Rodes*. From our present perspective, we could argue that Dalianidis used the *West Side Story* plot twenty years later for a number of reasons, for example, because the film's well-known plot and impact were so huge that he would otherwise have had to pay royalties to the producers or because he was heavily engaged with other films during his collaboration with FINOS FILM, the biggest Greek film production company in the 1960s. In the 1980s, Dalianidis could more easily erase the predominant "traces" of the archetype and, in the decade's postmodern spirit, allow the blending of his youth delinquency films with other genres, such as comedy and musical.

Apart from their elasticity, transnational adaptations may challenge the conventions and stereotypes of the "original." In this light, taking into account Davis's categorization, "contamination" doesn't take place unconditionally but conforms to formal, narrative, and iconographic standards of a local film industry. For example, the Greek film industry's close ties with American film imports shortly after World War II can explain direct transplants and contamination of Greek films of the period. However, there are instances, mainly regarding form, when adaptations "infiltrated"—that is, creatively adapted a dominant archetype and created an implicitly new version—popular Greek cinema, as locality was still considered a crucial factor in attracting audiences.

Cultural specificity usually prevails, as it exemplifies particular tensions within a local film industry and, by extension, society. In the case of *Rodes*, both *West Side Story* gangs were transcribed into Greek reality, echoing relevant social and cultural issues,[39] even though gangs were not common in the urban centers of Athens or Thessaloniki. Issues of juvenile delinquency were the growing concern of the day, as well as a further penetration of American trends in Greek mass culture by means of music, dancing, and clothing within a fluid mediatic terrain, which would be upheaved by the emergence of the local direct-to-video circuit from 1985 to 1990 and the deregulation of television after 1989.

It is no secret that transnational adaptations naturally preserve the main plot and the characters of the archetype, yet they tend to deconstruct and adapt it creatively to help the local/domestic audience construct meanings either in earnest or in a more cynical spirit. Of course, what is also really important is that the American archetypes are constructed to be adaptable to begin with. Therefore, it is tempting for a local film industry to engage in transnational adaptations. Perhaps we could argue in favor of "partial localization":[40] even though there is a dominant (US) film industry that distributes its films worldwide, only the most popular of the films are localized. This could be the case due to distribution as "a pre-condition for localization,"[41] but also because such transnational adaptations operate as "sets of metaphors by which people live as they help to constitute narratives of the other and protonarratives of possible lives, fantasies that could become the prolegomena to the desire for acquisition and movement."[42] So, apart from the usual framing of the United States' worldwide impact, it is important to examine how these localized film versions operate in the collective fantasies of their spectators.

To conclude, Smith's model of proliferation/industry/reception has proven to be most useful in our examination of transnational adaptations and the analysis of their relations with a dominant film archetype. However, one should also consider the very special conditions within a specific society, that is, the level of its "Americanization," the rise and popularity of particular film genres and their tradition, the sociocultural context, the time frame of the adaptation—especially when a vernacular version is adapted after several decades—as well as the plasticity of the archetypical timeless "mediascape" and the ways by which a filmmaker attempts to adapt it by "localizing" it. Furthermore, these elements could also shape one's understanding about the way an American genre becomes domestic and how it is embraced by the audience as such, erasing sometimes, but not always, all signs of the film industry from which it has proliferated.

# Notes

1. Arjun Appadurai, *Modernity at Large: Cultural Dimensions of Globalization* (Minneapolis: University of Minnesota Press, 1996), 35.

2. Ibid., 14.

3. See Darrell William Davis, "Reigniting Japanese Tradition with 'Hana-Bi,'" *Cinema Journal* 40, no. 4 (2001): 55–80.

4. Appadurai, *Modernity at Large*.

5. Anthony D. Pym, *The Moving Text: Localization, Translation, and Distribution* (Amsterdam: John Benjamins Publishing, 2004).

6. Iain Robert Smith, *The Hollywood Meme: Transnational Adaptations in World Cinema* (Edinburgh: Edinburgh University Press, 2017).

7. See, for example, Zygmunt Bauman, *Globalization: The Human Consequences* (Cambridge: Polity, 1998); Jürgen Osterhammel and Niels P. Petersson, *Globalization: A Short History* (New Brunswick, NJ: Princeton University Press, 2009); and Stephen D. King, *Grave New World: The End of Globalization, the Return of History* (London: Yale University Press, 2017).

8. Theodor Adorno and Max Horkheimer, *Dialectic of Enlightenment* (Stanford, CA: Stanford University Press, 2002).

9. John B. Thompson, *The Media and Modernity: A Social Theory of the Media* (Stanford, CA: Stanford University Press, 1995), 172.

10. Appadurai, *Modernity at Large*, 35.

11. Edward Said, *Culture and Imperialism* (New York: Knopf, 1993), 375.

12. Simone Murray, "The Business of Adaptation: Reading the Market," in *A Companion to Literature, Film, and Adaptation*, ed. Deborah Cartmell (Malden, MA: Wiley-Blackwell, 2012), 127.

13. Pym, *Moving Text*, 30.

14. Appadurai, *Modernity at Large*, 32.

15. Michele Hilmes, "The Whole World's Unlikely Heroine: Ugly Betty as Transnational Phenomenon," in *Reading Ugly Betty: TV's Betty Goes Global*, ed. Janet McCabe and Kim Akass (London: I. B. Tauris, 2013).

16. Davis, "Reigniting Japanese Tradition," 62.

17. Ibid.

18. Ibid., 65.

19. Ibid.

20. Smith, *Hollywood Meme*, 31–33.

21. Maria Paradeisi, "Erga kai Imerai tou Paliou kai tou Neou Ellinikou Kinimatografou" / "Works of and Old and New Greek Cinema," *O Politis* 122 (1993): 50–56.

22. Eliza-Anna Delveroudi, *Oi Neoi stis Komodies tou Ellinikou Kinimatografou 1948–1974 / Youth in Greek Cinema Comedies 1948–1974* (Athens: IAEN/INE, 2004).

23. Orsalia-Eleni Kassaveti, *I Elliniki Videotainia (1985–1990): Eidologikes, Koinonikes & Politismikes Diastaseis / The Greek Video-Film (1985–1990): Generic, Social and Cultural Aspects* (Athens: Asini, 2014).

24. Lydia Papadimitriou, *To Elliniko Kinimatografiko Musical / Greek Film Musical* (Athens: Papazisi, 2009).

25. At the same time, the trend of emergent consumerism (Kostis Kornetis, *Children of the Dictatorship: Student Resistance, Cultural Politics and the "Long 1960s" in Greece* [New York: Berghahn, 2013]) and the rise of tourism during the 1960s (Michalis Nikolakakis, "Tourism, Body and Seaside Recreational Practices in Postwar Greek Society until 1974," in *Consumption & Gender in Southern Europe since the Long 1960s*, ed. Kostis Kornetis, Eleni Kotsovili, and Nikos Papadogiannis [London: Bloomsbury, 2016], 103–17) operated as a means of financial empowerment for Greek society and were also reflected in some of the musical productions in the 1960s. Despite its popularity during the 1960s, the Greek film musical production faced decline alongside other genres due to the advent of television later in the same decade. In the following years, film musicals gradually disappeared, until their brief reemergence in the 1980s.

26. Nicos P. Mouzelis, *Modern Greece: Facets of Underdevelopment* (London: MacMillan, 1978), 120.

27. Chrysanthi Sotiropoulou, *Elliniki kinimatografia 1965–1975 / Greek Cinema 1965–1975* (Athens: Themelio, 1989).

28. Ibid., 45–46.

29. Whereas European films were previously sold on a prearranged fixed price, paid directly to the production company without any further commitment of the Greek distributor, the American films' commission of the proceeds was sent directly to the American distribution company, leading to a serious financial hemorrhage of the Greek film market.

30. Panos Kouanis, *I Kinimatografiki Agora stin Ellada (1944–1999) / Film Market in Greece (1944–1999)* (Athens: Finatec, 2001), 238.

31. Eliza-Anna Delveroudi, *Oi Neoi stis Komodies tou Ellinikou Kinimatografou 1948–1974 / Youth in Greek Cinema Comedies 1948–1974* (Athens: IAEN/INE, 2004).

32. The word designates the cinema of Turkey.

33. Orsalia-Eleni Kassaveti, "To Elliniko Melodrama" / "The Greek Melodrama," in *Apo ton Proimo ston Syghrono Elliniko Kinimatografo / From Early to Contemporary Greek Cinema*, ed. Maria Paradeisi and Afroditi Nikolaidou (Athens: Gutenberg, 2017).

34. Jane K. Cowan, *Dance and the Body Politic in Northern Greece* (Princeton: Princeton University Press, 1990), 4–5.

35. Kassaveti, *I Elliniki*.

36. See Grigoris Gizelis et al., *Paradosi kai Neoterikotita stis Politistikes Drastiriotites tis Ellinikis Oikogeneias / Tradition and Postmodernism in the Cultural Activities of the Greek Family* (Athens: EKKE, 1984), 127. According to their survey, the "Greeks believe in a traditional religious faith system," while they are considered "a very religious nation, as well as traditional in regards to their religious stance." The Greek Church has always tried to maintain a pivotal role in Greek society, especially during the 1980s.

37. Such musicals marked the revival of the old American musical in Hollywood, but in a modernized sense: they used prerecorded music (actors/actresses didn't sing), which could also be featured in video clips edited especially for commercial distribution, mainly on MTV. Furthermore, they shed a light on music subcultures, their cultural practices, as well as their fashion trends. Feuer describes these film musicals as "postclassical" with a MTV aesthetic in "Is Dirty Dancing a Musical, and Why Should It Matter?" in *The Time of Our Lives: "Dirty Dancing" and Popular Culture*, ed. Yannis Tzioumakis and Sian Lincoln (Detroit, MI: Wayne State University Press, 2013), 64.

38. Papadimitriou, *To Elliniko Kinimatografiko Musical*, 79.

39. Kassaveti, *I Elliniki Videotainia*.

40. Pym, *Moving Text*, 9.

41. Ibid., 14.

42. Appadurai, *Modernity at Large*, 36.

# 8

# "What'll Become of Me?"

## Adaptation and the Mediation
## of Black Women's Bodies

Nicole Pizarro

In the episode "Unbowed, Unbent, Unbroken" of the sixth season of *Game of Thrones* (HBO, 2011–18), fans were shocked to witness Sansa Stark (Sophie Turner) being raped on her wedding night by her husband, Ramsay Bolton (Iwan Rheon), while forcing Theon Greyjoy (Alfie Allen)—Sansa's childhood friend—to watch the attack. There were a few reasons behind the fans' reaction. In part, there was no narrative justification to the attack other than emphasizing Bolton's cruelty and his power over Sansa and Theon alike. In terms of cinematography, while the rape is happening, the camera focuses on Theon's face, as we hear Sansa's groans. What's more, this rape scene isn't featured in the *Game of Throne* book series. Although Bolton is a cruel character who rapes many women he comes across in the books, he doesn't rape Sansa. So why add this scene if it isn't in the source text?

Rape scenes throughout film history are not rare. In fact, rape and revenge film rose to popularity in the 1970s. In these films, rape was used as a narrative strategy and positioned the woman's violation as the main driving force for the film's plot. These attacks literally propelled the story forward, unlike episodes like "Unbowed, Unbent, Unbroken," where the attack has no basis. Nevertheless, the narrative tension already created was resolved during *Game of Thrones*'s penultimate episode of season 6, where Bolton suffers an agonizing death masterminded by Sansa. The same thing can't be said for the depiction of rape scenes in recent adaptations of slave narratives.

Slave narratives are a staple of the American literary canon and consist of biographical accounts from the point of view of freed slaves. Nonfictional autobiographical slave narratives, like Frederick Douglass's *Narrative in the Life of Frederick Douglass* (1845) and Harriet Jacobs's *Incidents in the Life of a Slave Girl* (1861), often trace the author's childhood, their experience as slaves, and their eventual escape from slavery. As raw accounts of American history, some slave narratives have been adapted onscreen, depicting the struggles that black bodies endured. In this chapter, I examine the relationship between adaptation and depictions of African American women in contemporary adaptations of slave narratives. I focus specifically on two films directed and written by black men who adapt slave narratives about or written by black men because these works demonstrate that black men directors perpetuate problematic representations of black women. I look specifically at Steve McQueen's *12 Years a Slave* (2013) and Nate Parker's *The Birth of a Nation* (2016), as both films were lauded by film critics[1] and both portray rape scenes that are absent in their respective source texts. I find this decision problematic, as it reinforces the position of black women as objects to be gazed upon and to be violated by white men.

Furthermore, the rape scenes problematize the position of the black woman spectator as they violate or penetrate her identification with the black bodies onscreen. In both narratives, the men who take up the role of these women's protectors leave them behind. *12 Years a Slave*'s Solomon Northup (Chiwetel Ejiofor) is freed from slavery and promptly abandons Patsey (Lupita Nyong'o) to her fate. Similarly, Nat Turner (Nate Parker) and his fellow slaves in the plantation he serves rebel against their owners but leave their wives, mothers, and sisters behind. The last time we see Patsey, she passes out when Northup forsakes her. Likewise, the last time we see Cherry Turner (Aja Naomi King) in *The Birth of a Nation*, she is still working for her owner. The women have no way out, and, by extension, the black woman spectator is also left without a way out.

## Adaptation Theory and the Representation of Black Women in Film

In *A Theory of Adaptation*, Linda Hutcheon makes a distinction between telling versus showing when it comes to adaptations. "The move from a telling to a showing mode may also mean a change in genre as well as medium, and with that too comes a shift in the expectations of the audiences."[2] Furthermore, Hutcheon argues that when it comes to adaptations, memory and the desire for repetition take hold, and she contends that adaptation "creates the doubled pleasure of the palimpsest: more than one text is experienced—and knowingly so."[3] This pleasure comes from repetition. But what happens when what is being told and shown is a traumatic experience? I argue that not only does adaptation facilitate pleasure through repetition, but it also facilitates trauma and violence through its transmedial nature.

I maintain that in adaptations of slave narratives, the black woman's body works as a site or locus of cultural trauma. When the black woman's body is assaulted, the violation is then transferred to the black woman spectator, and, ultimately, there can be no redemption, because, as shown in both case studies discussed below, the women are deserted by the men who are telling the story. Thus, as a practice that has the potential to extend its source text, adaptation becomes complicit in this assault. In "Beyond Fidelity: The Dialogics of Adaptation," Robert Stam contends that adaptations should be thought of as originals, in a way. Stam posits that each adaptation is an original work, separate from its source, and, as such, it needs to be studied in ways other than simply through the lens of fidelity. "The fact that the shots have to be composed, lit, and edited in a certain way . . . generates a difference [between the adaptation and its source text]."[4] Although adaptations have the space to deviate from their source text, a deflection such as including a sexual assault scene that doesn't get resolved raises the question of the extent to which adaptations should be held accountable for perpetuating harmful representations of black women on film.

When it comes to black spectatorship, bell hooks argues that independent black cinema was borne out of (white) cinema's "negation of Black representation."[5] Furthermore, hooks calls attention to how "major, early, Black, male, independent filmmakers represented Black women in their films as objects of the male gaze," resulting in "the Black male gaze [having] a different scope from that of the Black female."[6] Black women spectators partake in what hooks calls "the oppositional gaze," a sort of refusal of black women to identify with the desired white body on the screen. Contemporary African American cinema—or black cinema—came to fruition in the 1970s as a response to "a particular brand of Hollywood cinema" that capitalized on "films, starring Black actors, but produced by Whites and mostly directed by Whites [that] targeted Black audiences [and came to be known as Blaxploitation films]."[7] As Sheril D. Antonio asserts, "This decade clearly and distinctly marked the beginning of the contemporary period of African American cinema."[8] By the early 1990s, directors like Spike Lee, Mario Van Peebles, and Robert Townsend made a name for themselves within the industry. Today, black cinema is well established both in mainstream and independent circles. However, black filmmakers largely face the issue of lack of funding for their projects. As Monica Ndounou explains,

> Hollywood studio executives insist that a lack of international demand drives their investment choices and overall reluctance to distribute or produce films about women and African Americans. . . . Regardless of genre, [films with predominantly black casts] are perceived as niche films with limited market appeal. Such perceptions affect film conception, investment decisions, production, marketing, and distribution.[9]

One of the ways in which black filmmakers can secure funding for their works is by adapting slave narratives. When determining reasons for *why* filmmakers adapt certain works, Hutcheon[10] identifies several factors, including economy, cultural capital, legal constraints, and personal/political motives. Hutcheon argues that adaptations are safe bets when it comes to profit because they have preestablished audiences. This also applies to slave narratives, which not only have a place in the American literary canon but transcend the "niche" that most African American films find themselves in. Furthermore, slave narratives have cultural capital, which increases their film adaptations' appeal. That said, even with secured funding, slave narratives can still flop. In fact, "Black writers' explorations of slavery have historically been more accepted and profitable in print than on the big screen"[11] because "cinematic and other techniques that cater to crossover audiences frequently end up corrupting or eroding the positive development of African American film's form and content."[12] This also raises the question of cultural trauma and how certain films, if not carefully considered, can trigger it, as is the case of *12 Years a Slave* and *The Birth of a Nation*, which I will address in a later section.

Writing on theater, Peggy Phelan indicates that "performance implicates the real through the presence of living bodies. In performance art spectatorship there is an element of consumption: there are no left-overs, the gazing spectator must try to take everything in."[13] I claim that even though cinema and theater are two different mediums, the fact that they are both multisensorial and multimodal means that cinema also has the potential to complicate consumption, especially for bodies of color.

Although all bodies represented onscreen are sites of spectatorship and consumption, I claim that the relationship between the spectator and the gaze is complicated when both the gaze and the spectator belong to a disenfranchised community. In "'She Will Never Look': Film Spectatorship, Black Feminism and Scary Subjectivities," Terri Francis argues that "Black women's traumatic collective history . . . constitute[s] the background to Black women writers' and media artists' aesthetics."[14] The problem is that when black women's stories are told by black men, that collective trauma is emphasized. In the following sections, I analyze two scenes of rape in *12 Years a Slave* and *The Birth of a Nation* and attempt to make connections between their narrative purpose in the larger plot of both films. Later, I examine whether adaptations should be held accountable for this growing trend in film and television.

## 12 Years a Slave

*12 Years a Slave* tells the story of Solomon Northup, a freeborn African American man who was kidnapped and sold as a slave. Northup wrote and published his memoir in 1853, where he chronicles the events that led to his kidnapping, his subsequent twelve years in slavery, and his eventual freedom. Steve McQueen directed the film adaptation of *12 Years a Slave*, which was released in theaters in 2013. There are several discrepancies between Northup's memoir and its adaptation, but perhaps the most significant is the development of Patsey's storyline. Portrayed by Lupita Nyong'o, Patsey is a slave in Edwin Epps's (Michael Fassbender) plantation. In the film, we first see Patsey along with a group of slaves listening to their owner's sermon about obedience: "And that servant which knew his lord's will and prepared not himself, neither did according to his will, shall be beaten with many stripes." As he talks, the camera is focused on Epps in a close-up. However, as the camera pans by Epps's face, we see his wife standing behind him, at a higher angle though out of focus. Her quiet presence behind her husband, looking over his shoulder as he preaches, symbolizes the control that she holds over the slaves of the plantation, which is ultimately overshadowed by the fact that her husband has the final say on what happens at the plantation.

After Epps's sermon, the scene cuts to Patsey picking cotton along with Northup and the rest of the slaves. Patsey's skill in picking cotton results in Epps naming her the "Queen of the Fields," a "gift" that God bestowed upon him. In his memoir, Northup himself is the one who calls Patsey the epithet. He describes her as a "splendid animal and were it not that bondage had enshrouded her intellect in utter and everlasting darkness, would have been chief among ten thousand of her people."[15] However, Northup contrasts this description with the fact that despite Patsey's obedience to Epps, she "wept oftener, and suffered more, than any of her companions" because "it had fallen to her lot to be the slave of a licentious master and a jealous mistress." In the chapter titled "Patsey's Brows," Northup writes that Patsey

> walked under a cloud. If she uttered a word in opposition to her master's will, the lash was resorted to at once, to bring her to sub-jection; if she was not watchful when about her cabin, or when walking in the yard, a billet of wood, or a broken bottle perhaps, hurled from her mistress' hand, would smite her unexpectedly in the face. The enslaved victim of lust and hate, Patsey had no comfort of her life.[16]

Northup details how Epps would resort to whipping Patsey fiercely to appease his wife's anger. Mistress Epps hated Patsey and wanted Mr. Epps to sell her, but he refused, partly because she was the best cotton picker and partly because of his lust. In fact, in one particular instance in the memoir, Northup relates how, while hoeing next to Patsey, she began to cry, for she recognized "his lewd intentions."[17] This last instance doesn't result in Patsey being sexually assaulted, thanks to Northup's intervention. However, in the film, Patsey has no escape.

In the film adaptation, presumably the same night after this incident recounted in the literary version, Epps makes his way to the slaves' quarters and seeks out Patsey. The scene cuts to a two-shot of Epps on top of Patsey, who has a far-off look on her face. The only discernable sound is from crickets. Both bodies are silhouetted under the moonlight in a medium close-up. Epps keeps his eyes on Patsey, but she looks away and does not move while Epps assaults her. During the rape, the shot cuts to a close-up of Epps's face, and he stops and calls out Patsey's name. Then, the shot cuts to Patsey, lying motionless with Epps's hands around her throat. Epps slaps her and then, seemingly, ejaculates. After he finishes, a tear is visible going down his cheek. He puts his pants back on and walks away, leaving Patsey lying down, unmoving, like a rag doll.

When asked about the "scenes of perversion" in the film regarding Patsey, Epps, and Mistress Epps, McQueen claims:

> When people can do what they want and get away with it, they'll do it. The funny thing is, I have a lot of sympathy for Epps because he's in love with Patsey. He hates himself for it, and because he can't understand why he should love a slave, it comes out in the most vicious manner. We see him touching her tenderly, but then he rapes her. She switches off and he slaps her and tries to kill her. And then he wells up with emotion and he's in tears and walks away. No matter how people want to paint someone like Epps as evil or disgusting, he's a human being who can't escape his humanity.[18]

McQueen's depiction of Epps as a troubled, violent man doomed for his love toward Patsey raises the question of complicity in black men filmmakers' depictions of black women. If Spike Lee reproduces patriarchal patterns of filmmaking in his work, then McQueen is guilty of reinforcing those patterns and of forcing black women spectators to be the recipients of physical and emotional violence. Patsey's body is brutalized by everyone's hand. She is literally pushed to plead with Northup to murder her because she doesn't want to be alive anymore in a place where she is punished for existing and where obedience will never be enough.

Just like in Northup's memoir, Patsey is the recipient of many attacks from both Mr. and Mistress Epps in the film. In a scene shortly after Northup's arrival at Epps's plantation, Epps wakes up the slaves in the middle of the night and commands them to have a "party." The slaves are shown in their nightgowns, forced to dance to Epps's enjoyment. Patsey's dance moves catch his attention, prompting his wife to walk up to a cabinet, grab a heavy glass bottle, and throw it at Patsey's head. While Patsey screams in agony on the floor, Mistress Epps asks her husband to sell her, which he refuses to do, telling her, "Do not set yourself up against Patsey, my dear, because I will rid myself of you well before I do away with her." He then commands the slaves to keep dancing, and Patsey is seen being dragged away, crying.

As a character, Patsey is at a crossroads. It does not matter whom she is obedient to, because she'll still get beaten. McQueen's approach to Epps positions the plantation owner as a man who is unable to come to terms with the fact that he loves a slave, and the only way he can deal with his love for Patsey is through violence. Epps's wife, on the other hand, is unable to confront the fact that her husband loves and desires a black body more than her white body. Thus, she resorts to violating Patsey both emotionally and physically. This puts Patsey in a position that transcends both obedience *and* her positionality as a slave. Salamishah Tillet argues that "*12 Years a Slave* asks us to reconsider the implications of seeing *woman* and *slave* as mutually constitutive terms."[19] She argues that Northup uses Patsey in his memoir as a tool to further his abolitionist agenda. When he's done telling her brutal story, Patsey is absent from the rest of his narrative. "Ultimately, Patsey's absence . . . enable[s] Solomon to emerge as a thoroughly uncompromised hero at the expense of enslaved Black women."[20] This move serves, according to Tillet, to reinforce Patsey's role as representative of slave women in McQueen's film. However, *12 Years a Slave* isn't the only slave narrative adaptation to brutalize slave women's bodies to no narrative end.

## The Birth of a Nation

*The Birth of a Nation* tells the story of Nat Turner, who led a slave rebellion in Virginia. The events that led to Turner's uprising are cataloged in *The Confessions of Nat Turner, the Leader of the Late Insurrection in Southampton*. According to the text, God sent Turner a sign that he should revolt:

> And on the 12th of May, 1828, I heard a loud noise in the heavens,
> and the Spirit instantly appeared to me and said the Serpent was
> loosened, and Christ had laid down the yoke he had borne for
> the sins of men, and that I should take it on and fight against the

Serpent, for the time was fast approaching when the first should be last and the last should be first.[21]

It is unclear whether the women in Turner's life had any influence on his uprising. In the film adaptation of Turner's life, however, they are central to the story's progression.

*The Birth of a Nation* depicts various scenes of direct violence against women. The first one is toward Turner's future wife, Cherry (Aja Naomi King). In this scene, Turner and his master, Samuel Turner (Armie Hammer), are watching a slave auction. Turner's attention is drawn by a woman slave, who shows signs of extreme malnourishment. The auctioneer describes her as a "comely wench" and "not a day over 18." The auctioneer then rips Cherry's clothes off to reveal her breasts. The camera then cuts to a shot of a white man, rubbing his hands near his genitalia. The next shot is of Turner's face seemingly noticing the man's demeanor. He then turns to Samuel Turner and tries to convince him to buy her. The auctioneer continues the auction, telling the white men to imagine taking Cherry home, cleaning her up, and then assaulting her. At the last minute, Samuel Turner decides to buy her. Nat Turner then asks his mother and grandmother to help clean Cherry up and look presentable. The next time we see Cherry, she is working as a handmaiden in Samuel Turner's house.

Although Cherry's auction isn't a direct scene of assault, the scene helps position the viewer in terms of how black women's bodies are being mediated and consumed, as she is considered not a woman but an object to be used mainly for sexual gratification. The white men in the auction are not suffering like Epps in *12 Years a Slave*, and they are not ashamed of their feelings or do not shy away from them. They assault Cherry with their male gaze. This scene also helps to position Nat Turner as Cherry's savior, not unlike Northup with Patsey. The main difference between Northup and Turner is that Northup was forced at times, both in the film and in his memoir, to whip his fellow slaves, a position Nat Turner never finds himself in.

The first rape scene in *The Birth of a Nation* takes place when Cherry (now Nat's wife) is getting water from a well. She is approached by three white men who ask her to show them her permit to step ten feet away from the tree line. One of the men tells her, "Either you're going to show me a pass or you're going to show me something else." The scene cuts to Nat Turner asking Samuel Turner for a pass to go visit his wife, who has been attacked. The next time we see

Cherry, her face is badly beaten and almost unrecognizable. Turner kneels by her and comforts her while Cherry tries to apologize to him. Shortly after this scene, Samuel Turner hosts a party in his house where one of the members in attendance touches a woman slave's leg. The scene cuts to the woman slave's husband telling another one of Samuel Turner's slaves he refuses to allow his wife to get assaulted by the white man. Unfortunately, although Nat Turner tries to intervene, the slave is forced to sleep with the white man. Nat Turner and the slave's husband wait for her to come out, and they console her as she sobs.

## Rape as Narrative Strategy

Neither rape scene in *The Birth of a Nation* is recollected in Nat Turner's *Confessions*. One must recognize that even if Turner had told his white publisher Thomas R. Gray about the attacks, Gray would probably not have recorded them, as they would besmirch the names of slave owners. Furthermore, the rape scenes could represent historical assaults on black women by white men. However, taking into account that there are no historical documents that indicate that Cherry was ever physically harmed, the question remains as to *why* it was necessary to include such attacks in the film. One could argue that they certainly serve as narrative strategies to move the plot forward. Indeed, in *The Birth of a Nation*, it is through the women's battered, bruised, and raped bodies that the men find the courage to rise up against their oppressors. Yet, the women barely get a say in terms of how we read their bodies, as neither speaks after their attacks. Cherry's lack of a voice is physical, due to her bruised face. However, the other woman who gets raped never speaks. This reading is complicated by two details. First, why use rape as a narrative strategy to move the plot forward? Second, if the rapes serve as catalysts for the rebellion, why not show the women act?

Using rape as a narrative strategy is careless and a brutal affront to representations of black women on film. If the woman's body has diachronically been used as an "imaginary locus in which [audiences] could stage their anxieties about living in a world in which rape was a daily reality,"[22] then the violence gets doubly enacted upon the black enslaved woman's body. Not only is her body the display upon which the male gaze gets posited, but it is also visibly violated. Using rape as the vehicle for the progression of the film, especially when such a scene doesn't take place in the source text, sets a dangerous precedent for future adaptations of slave narratives. It might also be indicative of a larger trend in Hollywood. Men are the ones who adapt these texts, they write the screenplays, and they direct these films.

Ron Eyerman claims that "whether or not [African Americans] directly experienced slavery or even had ancestors who did, Blacks in the United States were identified and came to identify themselves through the memory and representation of slavery."[23] Thus, regardless of whether or not the physical act of penetration is presented to the spectator, black women spectators identify themselves through that act of violence. Women slaves were subjected to sexual assault constantly during captivity. Not showing the act doesn't necessarily take away from the collective memory of violence against black women's bodies that is ever present when we revisit slave narratives. However, adapting the text in a multimodal and multisensorial format multiplies that assault. Through the adaptation, the black woman spectator revisits that cultural memory and experiences the trauma of the assault by proxy of the performance.

In Northup's memoir, the last person he talks to before leaving Epps's plantation is Patsey, who throws herself on him and tells him, "You've saved me a good many whippins, Platt; I'm glad you're goin' to be free—but oh! de Lord, de Lord! what'll become of me?"[24] Immediately afterward, Northup disengages from her, gets in a carriage, and leaves the plantation. As he leaves, he turns around and sees Patsey fall to the ground. That is the last time Northup makes any mention of Patsey. In the film, no words are exchanged between them. They embrace, and Northup promptly gets on a carriage. He turns around once to wave at Patsey and doesn't look back. The shot of Northup leaving the plantation focuses on his face, and behind him we see Patsey's silhouette fall to the ground. Whereas in the memoir, Patsey voices her concerns toward Northup, in the film, we are haunted by her image because we have experienced the horrors forced upon her body by her oppressors. Ultimately, it is Northup's story that gets a resolution. We never find out what happened to Patsey in either tale. In the film, we don't even see her in focus as she falls to the ground. She becomes just another totalizing image of women slaves. Another silhouette that has to accept the fact that she will never be free.

We last see Cherry in juxtaposed shots between her, Nat Turner being hung after being found guilty for the uprising, and the black and white spectators. The crowds of black spectators watch in silence, as the white ones cheer for

Turner's demise. Cherry is shown hanging white sheets on the clothesline outside her owner's house. Although Turner eventually dies, he dies in arguable freedom. Not only is his wife left to carry the burden of slavery, but she is doubly condemned for being his wife. After the film ends, a title screen claims that after Turner's death, white people got scared of slaves and started slaughtering them for fear of another revolt. Even though the film prompts the black women as catalysts for the rebellion, they are eventually left in an even more precarious situation after the men—their husbands, brothers, and sons—get killed. Their own lives are at stake. Much like with Patsey, Cherry also becomes a locus of representation, and both black female characters continue to carry the burden of their race and their sex.

In both cases, there is a lack of consummation of the representation. By this, I imply that as images that black women spectators identify with, neither Patsey nor Cherry offers any sort of narrative resolution. Both women are left behind by the men driving the story, doomed to a life of slavery, whereas both Nat Turner and Northup achieve some sort of liberation—either through actual liberation or death. There is no concern that they will go back to being slaves. In fact, those spectators who identify with Turner and Northup already know they will have "narrative release," *because* the stories are adaptations of well-known "victory" narratives. If a significant percentage of viewers go to the theater knowing they are watching film adaptations, it would be fair to assume they expect Northup to receive his free papers and Turner to die by hanging. On the other hand, Patsey and Cherry come as a surprise. Indeed, Patsey does appear in Northup's tale, but the depiction of her rape does not. We are not expecting to see her body subjected to sexual assault. The same applies to Cherry, as well as the slave who gets raped but never speaks in *The Birth of a Nation*. Three women are attacked in both films to help the plot of the film move forward. By extension, the black woman spectator gets left behind.

In an analysis of a rape scene in Jonathan Kaplan's *The Accused*, Tanya Horeck[25] examines the ethos of representing rape in cinema and the implications of "looking at that rape." The film adapts a famous rape case in the United States, where a woman was violated by more than one man on a pinball machine, while other men watched. Horeck's study mainly questions the ethical implications of spectatorship in terms of the culpability of those men who witnessed the crime and said nothing and how that guilt almost transfers to the spectator. Ultimately, she questions "what happens when rape is turned into sensational public entertainment."[26] In this chapter, the problem is whether it is possible for rape to be used as a plot device when it comes to historical adaptations *because* this sort of attack took place historically speaking. How does historical accuracy figure in the pitfalls of the representations of rape in cinema?

## Adaptation and Cultural Trauma

Eyerman defines cultural trauma as a process "mediated through various forms of representation and linked to the formation of collective identity and the reworking of collective memory."[27] In the case of African Americans, slavery was the trigger for the cultural trauma. Because the slaves experienced trauma collectively, their identity was also collective in nature. In addition, "how slavery was represented in literature, music, the plastic arts and later, film, is crucial to the formation and reworking of collective memory and collective identity by the generations which followed emancipation."[28] Thus, the black body and its performance onscreen carry the weight of this trauma. However, whereas Hortense J. Spillers argues that scars on the flesh of the enslaved body "render a kind of hieroglyphics of the flesh" that "'transfers from one generation to another,'"[29] in *12 Years a Slave* and *The Birth of a Nation*, both the black body *in* the film and the black body of the spectator are the ones doing the act of remembrance. I argue that cultural trauma affects how the black female spectator relates to these adaptations' burden of historical accuracy.

For this discussion, it is imperative to consider what the implications of historical trauma are in adaptations of slave narratives. For that, I suggest looking into the psychic concept of adaptation proposed by Kamilla Elliott, who claims that in the process of adaptation "what should ideally pass from a book to film [is] 'the spirit of the text.'"[30] Elliott further diagrams what the psychic concept of adaptation looks like:

The novel's spirit → (The novel's form) → (Filmmaker Response) → (Film) → Viewer Response[31]

It follows that the novel's spirit always remains the same and exists as separate from the actual novel. The spirit of the novel is always present; it does not depend on the novel's form—or any mediation. This concept is used in adaptation studies when talking about the issue of fidelity. The adapter aims for the spirit of the novel and not necessarily the novel's form when adapting. However, this model suggests a homogeneity of the audience responding to the adaptation. I argue that, by parting from Elliott's diagram, it is possible to theorize what history and trauma do to a *nonfiction* novel's spirit and how that affects the viewer response. The diagram would look like this:

[History (Trauma)] → The [memoir's] spirit → (The [memoir's] form) → (Filmmaker Response) → (Film) → Viewer Response [Cultural Trauma]

It is impossible to separate any nonfiction text from the history and the trauma that might or might not be associated with the portion of history the text takes place in. They affect the novel's spirit. Therefore, it is arguable that the spirit of a novel based on facts and memory—as is the case of slave narratives—can't exist without the trauma. At the same time, then, cultural trauma is ever present, and it is tied to the viewer response. Ndounou contends that "cultural trauma significantly influences representations of history in all media" and that "the trauma or revisiting stories about slavery and other horrors can trigger unresolved, negative emotions."[32] If history and trauma are always present in slave narratives, then it is understandable why the rape scenes were included in *12 Years a Slave* and *The Birth of a Nation*. As adaptations, both films attempt to remediate the source texts' spirits. However, this raises the question of whether adaptations *should* allow for this sort of representation. Furthermore, are black male filmmakers complicit in this violence?

In "Reading Through the Text: The Black Woman as Audience," Jacqueline Bobo posits that in Steven Spielberg's adaptation of *The Color Purple* (1985) there is a shift in the emphasis of the story from the black woman to the black man. She argues that this is the result of Spielberg, a white man, taking up the task of this particular adaptation.[33]

Furthermore, Bobo contends that literary works such as *Incidents in the Life of a Slave Girl* (1861) and *The Color Purple* (1982) were written to counter "the pervasive depictions of the sexually promiscuous Black woman"[34] and that Spielberg, in his adaptation of Alice Walker's novel, reinforces negative stereotypes of black womanhood:

In paying good attention to a film in which aspects of their histories were depicted, Black women were able to extract images of power and relate them to their lives. In the best of all possible worlds, Black women would be able to see themselves represented in films that are directed by Black women. Unfortunately, there are no Black women who are allowed (meaning financed) to mount the kind of production as that given [to films such as] *The Color Purple*.[35]

Bobo adds that the film adaptation of *The Color Purple* would have been more successful in terms of how the story is portrayed had it been made by a black woman, especially because, as a white man, Spielberg frames the adaptation of the novel in accordance to his "experiences, cultural background, and social and political worldview."[36] Thus, his adaptation ends up being less concerned about race than its source text, which disregards Walker's positionality as a black woman.

Black male filmmakers are complicit in violent representations of black women's bodies. Jacquie Jones compares black films directed by black male directors as "overidentified with White maleness and the systems of sexual oppression in the society at large."[37] In "The Construction of Black Sexuality," Jones traces how Spike Lee presents black women in his films as "powerless and, in one way or another, abused."[38] Furthermore, she contends that black cinema integrates sexual themes in the main action of black films. The problem arises when this integration is constantly violent, as is the case of film adaptations of slave narratives. However, Ndounou argues that "[black] filmmakers are pressured to abide by the unwritten rule that a film must incorporate a white point of entry regardless of historical time period or context."[39] If black male filmmakers reinforce "white" ways of seeing, while at the same time making "sex" central to the plots of their films, there is no escape for black women spectators' bodies to escape the violence.

# Conclusion

The surge of rape scenes in popular film and television shows is dangerous. In *Game of Thrones*, we experience Sansa's rape at the hands of her husband. In terms of representation of women in film, the brutalization of female bodies onscreen is problematic—especially when there is no narrative resolution for the scenes in terms of the overall plots of the stories or, as is the case with the adapted texts of this chapter, when these violent acts are not included in the literary source. The rape scenes either serve no purpose or only move the man's story forward. In *The Birth of a Nation*, the camera cuts before the assault take place. We do, however, see both of the black women's bruised and battered bodies after the attacks. In *12 Years a Slave*, the audience is forced to sit through the act. In the case of black representation, cultural trauma itself might be the reason behind these scenes.

Using rape as narrative strategy in these adaptations may be black male filmmakers' way of avoiding cultural amnesia, which "persists when [black] perspectives on slavery are not treated as worthy of cinematic exploration."[40] Including the rape scenes in the film adaptations cements the impossibility of separating history from the spirit of the novel and, therefore, the film adaptation and the viewer's response. The avoidance of cultural amnesia has the potential of resulting in "legalizing" the brutalization of black women's bodies onscreen. When it comes to using rape as a narrative strategy, there are no right or wrong answers. One thing is certain, however: black cinema is constantly evolving, and so are the ways of looking at black bodies in popular culture.

# Notes

1. See, among others, Peter Travers, "12 Years a Slave," *Rolling Stone*, October 17, 2013, https://www.rollingstone.com/movies/movie-reviews/12-years-a-slave -113891/; Susan Wloszczyna, "12 Years a Slave," *Rogerebert.com*, October 18, 2013, https://www.rogerebert.com/reviews/12-years-a-slave-2013; and A. O. Scott, "Review: In Nate Parker's 'The Birth of a Nation,' Must-See and Won't-See Collide," *New York Times*, October 6, 2016, https://www.nytimes.com/2016/10/07/movies/ the-birth-of-a-nation-review-nate-parker.html.

2. Linda Hutcheon, *A Theory of Adaptation*, 2nd ed. (Oxon: Routledge, 2013), 45.

3. Ibid., 116.

4. Robert Stam, "Beyond Fidelity: The Dialogics of Adaptation," in *Film Adaptation*, ed. James Naremore (New Brunswick, NJ: Rutgers University Press, 2000), 56.

5. bell hooks, "The Oppositional Gaze: Black Female Spectators," in *Black American Cinema*, ed. Manthia Diawara (New York: Routledge, 1993), 289.

6. Ibid., 291.

7. Susan Hayward, *Key Concepts in Cinema Studies* (London: Routledge, 1996), 24.

8. Sheril D. Antonio, *Contemporary African American Cinema* (New York: Peter Lang, 2002), 5.

9. Monica W. Ndounou, *Shaping the Future of African American Films: Color-Coded Economics and the Story Behind the Numbers* (New Brunswick, NJ: Rutgers University Press, 2014), 4.

10. Hutcheon, *A Theory of Adaptation*.

11. Ndounou, *Shaping the Future*, 133.

12. Ibid., 132.

13. Peggy Phelan, *Unmarked: The Politics of Performance* (London: Routledge, 1993), 148.

14. Terri Francis, "'She Will Never Look': Film Spectatorship, Black Feminism and Scary Subjectivities," in *Reclaiming the Archive*, ed. Vicki Callahan (Detroit, MI: Wayne State University Press, 2011), 111.

15. Solomon Northup, *Twelve Years a Slave* (Baton Rouge: Louisiana State University Press, 1968), 188.

16. Ibid., 189.

17. Ibid., 227.

18. Quoted in Graham Fuller, "Of Human Bondage," *Film Comment* 49, no. 5 (2013): 20, MasterFILE Premier, EBSCOhost, 14.

19. Salamishah Tillet, "'I Got No Comfort in This Life': The Increasing Importance of Patsey in *12 Years a Slave*," *American Literary History* 26, no. 2 (2014): 355.

20. Ibid., 358.

21. Nat Turner and Thomas R. Gray, *The Confessions of Nat Turner: The Leader of the Late Insurrection in Southampton, Va.* (Chapel Hill: University of North Carolina Press, 2011), 11.

22. Kathryn Gravdal, *Ravishing Maidens: Writing Rape in Medieval French Literature and Law* (Philadelphia: University of Pennsylvania Press, 1991), 18.

23. Ron Eyerman, *Cultural Trauma: Slavery and the Formation of African American Identity* (Cambridge: Cambridge University Press, 2001), 14.

24. Northup, *Twelve Years a Slave*, 308.

25. Tanya Horeck, *Public Rape: Representing Violation in Fiction and Film* (London: Routledge, 2004), 91.

26. Ibid., 92.

27. Ron Eyerman, *Cultural Trauma: Slavery and the Formation of African American Identity* (Cambridge: Cambridge University Press, 2001), 1.

28. Ibid., 10.

29. Hortense J. Spillers, "Mama's Baby, Papa's Maybe: An American Grammar Book," *Diacritics*, 17, no. 2 (1987): 67.

30. Kamilla Elliott, "Literary Film Adaptation and the Form/Content Dilemma," in *Narratives Across Media*, ed. Marie-Laure Ryan (Nebraska: University of Nebraska Press, 2004), 222.

31. Ibid., 223.

32. Ndounou, *Shaping the Future*, 133.

33. Jacqueline Bobo, "Reading Through the Text: The Black Woman as Audience," in Diawara, *Black American Cinema*.

34. Ibid., 274.

35. Ibid., 285.

36. Ibid., 279.

37. Jacquie Jones, "The Construction of Black Sexuality: Towards Normalizing the Black Cinematic Experience," in Diawara, *Black American Cinema*, 254.

38. Ibid., 255.

39. Ndounou, *Shaping the Future*, 134.

40. Ibid., 137.

# 9

# Adaptations of Adulthood

## Toward a Model for Thematic Rhetorics in Adaptation Studies

Joakim Hermansson

THE REASON WE CONSUME fiction is rarely just to enjoy the superficial pleasure in escaping everyday life. On the contrary, Patrick Colm Hogan argues, it is the possibility to find keys for a somewhat happy life of normalcy and "enduring satisfaction" that makes us return again and again to fictional worlds and their characters.[1] For that reason, readers and spectators use fiction for mental simulations of life patterns, personal ethics, and paths for the future.[2] To present useful thematic lessons with rhetorical efficiency, award-winning and best-selling novels, art films, and Hollywood blockbusters make use of universal plot structures. For the knowing audience, the experience can be heightened further by adaptations' potential to offer "a constant oscillation" between two or more texts that relate, as equals, to a thematic problem.[3] As Lars Ellerström underscores, "a typical film adaptation, for instance, does not represent its source novel—it is not a film about a novel; rather, it represents the story, characters, and so forth that were earlier represented by the novel."[4] Hence, the oscillation is not restricted to a pair of versions when novel/film adaptations are concerned.

In one of many approaches that open the imagination for less limited relationships than one-on-one cases, Sarah Cardwell outlines the notion of an urtext, which is not the first artifact in a line of adaptations, but "a sort of 'myth,' an ur-text that stands outside and before each retelling of the story, and which contains the most fundamental parts of the tale without which an adaptation would lose its identity as that tale."[5] The various texts, such as novels and films that relate to a defined urtext, are "part and parcel of creating the larger web,"[6] and together, as nodes, they form a map of the urtext in question. In light of Hogan's reasoning, it is reasonable to consider the story of adult normalcy as the urtext of all fiction. People even employ thematic story structures to construct their life stories, form their senses of self, and make strategic choices in life,[7] and they thus regenerate variations of the urtext that tie most fiction to the same knot: the problems of becoming and being an adult human being.

In this chapter, I turn to ten successful contemporary novel/film adaptations to examine the rhetorical representation of the urtextual story about adulthood through the lens of Joseph Campbell's *A Hero with a Thousand Faces*.[8] The protagonists' thematic functions are imperative for the analyses, but it is the thematic and rhetorical progression, and content of the argument itself as it passes through the stories, that is my focus here. The inevitable consistencies and variations between any bodies of films and novels and their representations of adulthood mean that two often conflicting perspectives regarding adaptations must be maintained in tandem. If the novels are regarded as primary adaptations, the films can be thought of as secondary adaptations, dependent on the novels as source texts. Still, both versions should be seen as equals in an adaptation pair. Regardless, my sample movies form a body that represents film adaptations of the urtextual story about adulthood, and the novels form a corpus of its own. However, the purpose is not to address media specificity, or to produce statistical evidence for certain tendencies, but to sketch a map of the grounds for thematic reflection that films and novels have to offer together and to suggest a model for thematic studies of adaptations.

It is in this context that Campbell's presentation of the monomyth is instrumental. Examining myths around the globe and their representation of life's mixture of progression and obstacles, Campbell extracts a narrative model of the hero's journey, in seventeen developmental steps toward "adult realization,"[9] and a possible full "reintegration with society" as a responsible adult.[10] Campbell indicates that the steps of the hero's journey match a four-act plot structure. In the table below, I group the steps of the hero's journey accordingly and slightly adjust Campbell's sometimes obscure terminology for the sake of clarity.[11]

### *The rhetorical steps of the hero's journey in four acts*

| Act 1: Introduction | Act 2: Complication | Act 3: Development | Act 4: Conclusion |
| --- | --- | --- | --- |
| The Normal World | Road of Trials | Atonement | The World Denied |
| Call to Exploration | An Approach | Apotheosis | Flight and Rescue |
| Reluctance to Change | Temptation | Clarity | Return to the Normal World |
| Mentoring | | | Master of Two Worlds |
| Crossing the First Threshold | | | Freedom to Live |
| Belly of the Whale | | | |

My argument is that the hero's journey forms a line of argument about adulthood and a most useful tool for, but not limited to, thematic adaptation studies.[12] Still, for the hero's journey to be useful for thematic studies, its rhetoric progression needs to be complemented by a model that describes the thematic markers in question.

Adulthood has been theorized from many perspectives and can be regarded as an idealized vision of what it is to be a human being or as a long-winding road of trials. Based on several existing models, I have synthesized a model of seventeen thematic *markers of adulthood* in four groups.[13] The first group of markers describes structural aspects of life: financial autonomy, home, family, occupation, social network, and adult recognition. The second group concerns social facets: conformity, social adaptability and cooperation, care for others, and commitment in terms of intimacy and loyalty. The third group relates to substantive markers: stable values, self-awareness and self-esteem, emotional balance, and the awareness of complexities. The final group covers executive qualities: goal-oriented and rational thinking, responsible judgments and choices, and rational actions.[14]

In works of fiction, the respective significance of thematic markers is expressed and emphasized directly or indirectly through various modes of characterization, like narration, speech, thoughts, emotional reactions, and physical actions, that form rhetorical patterns through their distribution. Incidents and sections, which deviate from normal patterns of characters and narrations in a story, draw attention to various thematic markers.[15] For instance, attention can be drawn if a confident character hesitates or a narrator switches narrative style. Thus, characters are given thematic dimensions and functions to make a narrative's argument clearer.[16] During the course of a story, the focus may therefore shift to accentuate secondary characters' thematic dimensions, as a context for the central incidents and characters.[17] Furthermore, each step in the hero's journey is inherently associated with certain rhetorical and thematic qualities, so the position of an event in the plot affects its meaning, for example, awareness of complexities when a broader perspective is needed to achieve the goals.[18]

With these principles in mind, I coded the expressions that draw attention to markers of adulthood in ten novel/film adaptations to explore the representation of adulthood. To make the selection of titles less subjective, I employed a number of criteria. First, the main characters are of adult age. Also, both novels and films are published and distributed, respectively, in the English language during the twenty-first century. Moreover, all novels and films are commercially successful and/or critically acclaimed.[19] Finally, I have strived to include a wide spread of fiction genres and to attain a reasonable balance between male and female protagonists, authors, and (if possible) directors. This resulted in the following novel/film pairs: *Atonement* (2001,[20] 2008), *Fifty Shades of Grey* (2012,[21] 2015), *Gone Girl* (2012,[22] 2015), *Me Before You* (2012,[23] 2016), *Room* (2010,[24] 2016), *Shutter Island* (2003,[25] 2010), *The Da Vinci Code* (2003,[26] 2006), *The Martian* (2014,[27] 2015), *The Road* (2007,[28] 2009), and *Up in the Air* (2001,[29] 2010).

## Act 1: Setup

Beginning with a presentation of *The Normal World*, most novels' and films' first act frequently stage the storyworld's or the protagonists' conventional notion of adulthood, with a limited awareness of the narrow "familiar life horizon" and the "infantile ego" within.[30] This implies that there will be a display of shortcomings regarding the social markers of adulthood (i.e., rules and conventions, care for others, intimacy and loyalty, social adaptation and cooperation), as well as inadequate awareness of complexities, self-esteem, sense of responsibility, and rationality. Consequently, the first step in the narrative progression, *Call to Exploration*, points out the need for a less simplistic and self-centered approach to the world, for action and responsibility. The beginnings of the sample films and novels follow this design remarkably well, apart from minor variances.

In some cases, markers of adulthood are demonstrated as obvious short-comings; in *Me Before You*, Louisa's (in the film played by Emilia Clarke) limited worldview is expressed through her belief that every problem can be solved with a cup of tea.[31] In *Fifty Shades of Grey*, Christian (Jamie Dornan) declares that he doesn't "have a heart,"[32] while Ms. Steele (Dakota Johnson) claims to be "missing the need-a-boyfriend-gene,"[33] which also indicates a problem to relate to others due to some emotional imbalance. In other cases, the positive value of a trait is illustrated, such as the highlighted intimacy in both versions of *The Road* and *Room*. In the beginning of the screen version of *The Da Vinci Code*, Langdon's (Tom Hanks) lecture about perception and interpretation focuses entirely on the awareness of complexities in the social world.[34] Other common tendencies are the films' and novels' emphasis on having meaningful work, a home, emotional self-control, a developed sense of rationality, and a goal-oriented way of thinking as an adult. Nevertheless, most of the protagonists are concerned because they have no clear plans for the future.

Next, a *Reluctance to Change* demonstrates the strengths of the habitual concept of adulthood. In the exemplar stories, a status quo and the initial egocentric viewpoint are justified because simplifications make life more convenient. Many of the protagonists also take to lies to defend their ambivalence toward conformity, care for others, rationality, responsibility, and emotional balance. Also, the films primarily question the necessity to complicate matters of self-esteem, sociability, and cooperation, while the novels generally demonstrate the impracticality of a complex worldview at this stage. The novel *Up in the Air* summarizes the key argument for this convention with the answer from the protagonist, Ryan Bingham (George Clooney), to the question of what he is looking for: "Low maintenance."[35] With the same desire for constancy and simplicity, a frustrated Cecilia (Keira Knightley) in the film *Atonement* rudely dismisses and escapes her brother's question, and her own feeling that something has changed between her and her childhood friend Robbie, by diving into the pool.[36]

In response to the reluctance to change, the *Mentoring* phase draws attention to what can be gained from a development based on a more responsible, broad-minded, goal-oriented, and rational mindset. Accordingly, when *Shutter*

*Island*'s Dr. Naehring (Max von Sydow) characterizes the protagonist's nature, he also instructs the spectators that "retreat isn't something you [should] consider an option" at this stage.[37] *The Martian*'s protagonist Mark Whatney instead finds an inner mentoring voice as he asks what Hercule Poirot would do in his situation.[38] As a rule, at this phase of the journey, the novels again stress the urgency to consider the world's social complexities. Hence, in *Gone Girl*, a concerned Nick wonders warningly if "we are actually humans at this point."[39] To motivate a further thematic exploration, the films commonly illustrate the risk of stagnation and regression of adulthood, through scenes of people losing their jobs and self-respect.

What remains of the first rhetorical act is a final commitment to responsibility and change through a leap into the unknown, *Crossing the First Threshold*. This is represented by Langdon's "academic curiosity" in *The Da Vinci Code*,[40] Robbie's dream to serve humanity as a doctor in *Atonement*,[41] and Ms. Steele's plea, "enlighten me then," in *Fifty Shades of Grey*.[42] However, this step requires a taxing sacrifice of the ego, for the sake of community, care for others, and social adaptation.[43] So, just before Bingham is exposed to his own relative isolation in *Up in the Air*, his CEO (Jason Bateman) encouragingly tells him that he "will not be alone."[44] In the next scene, Bingham packs a huge cardboard poster of his sister and her husband, which he reluctantly has promised to bring with him, and then he sees his work partner being hugged goodbye by her boyfriend. The sequence also relates to the value of intimacy, loyalty, and responsibility as balanced parts in a cultivated concept of adulthood. By implication, the future prospects are at this stage associated with aspirations of home and family. In *Shutter Island*,[45] it is the contrast between the memory of a father who "was a stranger . . . to everyone" and the protagonist's own desire to be an emotionally balanced family man that takes the narrative across the threshold. Finally, the first act's rhetorical departure concludes with *The Belly of the Whale*, "a form of self-annihilation,"[46] as the old view on adulthood is left behind. "I feel I could disappear," *Gone Girl*'s Amy (Rosamund Pike) states to express the urgency to continue[47] and to foreshadow the necessity to accept normal life as an anonymous but functional part of society.

## Act 2: Complication

The second act comprises the first part of what Campbell refers to as "initiation."[48] Through tests and trials under new circumstances, it complicates the notion of adulthood, leading to a provisional attempt at closure and a temptation to either settle with the development, as far as it has progressed, or abandon it altogether. Christopher Vogler's name for the first part of the act, "Tests, Allies, Enemies,"[49] highlights its overall purpose to sort out the features that can serve as allies and obstacles for an advanced sense of adulthood. During this phase, the capacity to make responsible judgments and choices is underscored in the novels and films, and the characters are now made aware of "a whole heap of responsibility," to quote Louisa in *Me Before You*.[50] Thus, coping with the demands, the mother in *Room* is "trying to be realistic" as she takes inventory of ideas to escape from her captor.[51] Throughout the trials, goal orientation and rationality are also accentuated, as is the link between responsibility and cooperation. This point is stressed when *The Martian*'s NASA director (Jeff Daniels) senses the staff's faltering commitment to exhaust all coordinated efforts to rescue Whatney and comments that "Mark dies if you don't."[52] The increasing number of issues that the characters must handle brings forth the importance of care, guiding principles, stable emotional reactions, and a rising self-esteem. It is with those qualities in balance that Ms. Steele is capable of taking a stand and managing a negotiation with Christian in *Fifty Shades of Grey*.[53] Particularly, the films now reemphasize the value of home and family as a stable base for adult life, while loyalty under pressure is rather problematized. In *Atonement*, most of this phase is set at a family dinner, after which Cecilia fails to support Robbie against her family's joint accusations. In *The Road*, a flashback of the man's (Viggo Mortensen) previous family life, just when he leaves his wife behind, sets the tone.[54] As an alternative, the novels expose various traditional beliefs, guiding principles, and strict conformity to social conventions, or "protocol,"[55] as outdated and highly impractical flaws in the governing definition of adulthood. In *Me Before You*, Will's mother wrestles with her convictions to let him die,[56] while Will questions Louisa's dismissal of new experiences on the grounds that she is "not that sort of person."[57] Likewise, the man in *The Road* has to reevaluate what it is to be good, when he realizes that taking care of his son means that he is ready to kill, although he believes he is given the role as carer by God.[58] In *The Da Vinci*

*Code*, Langdon learns to balance "*right*eousness, dexterity, and correctness" with "irrational thought."[59]

The latter two steps of the second act assess what has been learned thus far. First, during *The Approach*, various markers are coupled in juxtaposed images, actions, and statements to present the "totality of what can be known"[60] through a "harmonization of all the pairs of opposites."[61] This draws immediate attention to the awareness of complexity as an adulthood marker, and some statements in the novels and films—such as "There's a very fine line between pleasure and pain. . . . They are two sides of the same coin"[62]—are demonstrative of the point. In *Atonement*, the contrast between human passion and logic is problematized when the scene with Robbie (James McAvoy), standing helpless in front of dead schoolgirls in wartime France, is juxtaposed with a flashback of Briony (Soarsie Ronan) as a child throwing herself into a pond to make Robbie prove his affection by saving her.

As the latter scene indicates, the focal point is self-awareness and self-esteem, and the sample films and novels now illuminate this marker's sophistication and the key role it plays for adulthood. Financial autonomy, work, a social context, cooperation, and care, as well as loyalty and intimacy, all build the characters' self-esteem, which in turn facilitates the internalization of the other attributes. This results in a better sense of judgment and courage to assume responsibility, we are told. At this point, the films tend to assert that mechanical conformity to conventions reduces a person's self-esteem and that strict obedience to them weakens the capacity to exercise intimacy and loyalty. In *Up in the Air*, Natalie (Anna Kendrick) realizes how she has diminished her own worth by picking a partner because "he fit the bill." To the same effect, in *Me Before You*, Louisa delivers an illustrative monologue about the self-reducing threat that conformity poses, through standardized family life, recommended careers, prescribed hobbies, and a focus on appearance. The attention on self-awareness and life's social intricacies also summons distinct expressions of principles and faith as a reaction. These are often combined with some disillusion. In *The Da Vinci Code*, a belief "in people" is expressed merely because "it is all we have," while the balance between fiscal costs and the value of science, in relation to the significance of a single human life, is addressed with the question "How much is too much?" in *The Martian*.[63]

All these complications illustrate the distinction between adulthood and "inappropriate sentimentalities and resentments [and] . . . childlike human convenience."[64] Moreover, there is a reevaluation of the competencies associated with childhood and the ability "to dream the dreams of a child's imaginings," as the man articulates in *The Road*[65] to amplify a sense of consciousness.[66] Consequently, Louisa sings the innocent "Molahonkey Song" from her childhood to Will, who needs to regain his spirit, in *Me Before You*. The rhetorical purpose is to integrate emotional presence with rationality and an expanded sense of responsibility to produce a new level of maturity.

The experience can, however, be overwhelming, since the process toward a developed thematic understanding has only reached its midpoint, with core issues still unresolved, as articulated in *Me Before You*: "Old emotions washing over me, new thoughts and ideas being pulled from me as if my perception itself were being stretched out of shape."[67] That the lesson, so far, is more abstract than practical invites the idea to settle or even to return to the original perception of adulthood. So, the next step, *Temptation*, tests the current "position of reference" to the "general human formula"[68] and takes stock of ideals and remaining conflicts. Attention is given to the fact that adults are expected to be able "to make tough decisions"[69] and to issues of beliefs and principles that may support them. Still, the narratives mainly return to human convenience and ego, through either the characters' doubts or their contentment. The boy in *Room* gives voice to the entire idea of retreat, as well as his mother's thoughts, in the proclamation: "I have seen the world and I'm tired now."[70] In other cases, it is the desire to have a family, home, and work that incites an impulse to withdraw. In *The Road*, this is accentuated through the man's memories of his wife "crossing the lawn,"[71] while he and the boy wish that they could stay in the comfortable and well-equipped homelike bunker they have found. With a touch of irony, *The Martian* wonders if yearning for the comforts of home is reason enough to leave Mars, where he can say he is "the greatest botanist on this planet."[72]

However, the individual's importance as part of a greater community creates a drive to go further in both films and novels, to "go outside and look at the vast horizon," as *The Martian* advises.[73] The novels also support the movement forward through an emphasis on inner convictions and a belief in what is true. "Liars, liars" is the cry to further deliberations that takes *Atonement* to the third act.[74] Similarly, *The Da Vinci Code* provides an argument to tip the balance with the declaration that "those who seek the truth are more than friends,"[75] an idea that will unite ideals and pragmatic politics at the end of the story.

# Act 3: Development

In the third act, *Development*, the struggle to complete the thematic grasp of adulthood, begins. This demands a major revision of a core issue, *Atonement*, which leads to *Apotheosis* and *Clarity* as rhetorical steps. Following the second act's distinction between constructive and corrupt thematic aspects, it is time to atone with the idea that they are part of a whole. Above all, the thematic notion must be detached from the egocentric perspective,[76] guided by inner values. It follows that the rhetorical concentration on self-awareness and self-esteem escalates at this point in both novels and films. Typically for the films, the characters' feeling of irrelevance develops into a potent sensation that they are "supposed to be happy," as phrased in *Me Before You*,[77] but cannot be so unless they serve a purpose and are needed. This challenge is formulated as a question in *Up in the Air*—"God is claiming people. Am I still on his list or has he skipped me?"[78]—and as an instruction in *Shutter Island*—If "you want to uncover the truth, you gotta let it go."[79] In *Gone Girl*, Amy transforms her appearance to become an anonymity,[80] and *Atonement*'s famous, elegiac long take depicts the individual's insignificance in the mass of a defeated army. The latter also emphasizes, like many of the films, a low-key intimacy of community and the value of faith, as the soldiers unite in groups to the sound of a hymn.[81] Gradually, with the world's revealed complexity forming a unified image, a redirection toward a greater care for others in life emerges.

This transformation of the mindset paves the way for emotional balance, or "peace at heart," which facilitates the unification of good and bad human qualities.[82] Thus, "no longer self-conscious"[83] and aware that she is the meaning of life for another person, Louisa dances inappropriately close to Will at a wedding in *Me Before You* and demonstrates that it is possible to reform conventions and display intimacy in public without fear or shame.[84] In *The Martian*, Rich Purnell (Donald Glover) also breaks with the expected behavior pattern and explains a complex space maneuver to NASA management the way a child would, pretending to be the spacecraft, while the others in the room adapt and accept his style. He concludes his demonstration stating he has "done the math. It checks out,"[85] as if to underscore the efficiency of including unconventional approaches in the repertoire.

The notion of adulthood is now developed enough to serve as guidance through the dilemmas and necessary prioritizations that its practice provokes.

The narratives demonstrate that, with a grounded self-esteem that also encourages nonconforming styles and manners, the adult is ready to become "dedicated to the whole of his society."[86] Accordingly, *The Da Vinci Code* tells the reader that the old "ideal that man must be told what to do . . . because man is incapable of thinking for himself" can be replaced with the rule that the adult can handle the "*truth* and be able to think for himself."[87] Correspondingly, Ms. Steele realizes in *Fifty Shades of Grey* that she might be the one to "guide [Christian] into the light" instead of the other way around.[88] However, this commitment comes with an inevitable obligation, expressed in both novels and films. *The Da Vinci Code* proclaims that you must "embrace this responsibility . . . or you must pass [it] . . . to someone else,"[89] and Teddy Daniels (Leonardo DiCaprio) in *Shutter Island* decides "to find out" the truth, although "it's a short step from suicide,"[90] indicating that the last act's resolution is near.

The third act ends with "the ultimate boon" of *Clarity*, a summary of the thematic comprehension as an "indestructible body,"[91] or "a thing that even death cannot undo," as *The Road* articulates.[92] "He felt so small, so utterly human, but it wasn't a debilitating feeling. It was an oddly proud one. To be a part of this. A speck, yes. But part of it, one with it," the narrator in *Shutter Island* explains.[93] In *Up in the Air* Bingham places a picture on a map of the United States at his sister's wedding, first disappointed to see his contribution vanishing among all the others but then smiling faintly with a growing realization that he is a part of a complex whole that does not fall apart into fragments but forms a substantial unity.[94] In that context, the films foreground care, cooperation, family, and home as vital components to solidify adulthood.


## Act 4: Conclusion

The fourth act is often regarded as "a straightforward progress toward the final resolution"[95] and a thematic conclusion. However, Campbell posits that it begins with a display of "obstruction and evasion,"[96] as the abstract insights and ideals must be implemented in the far from effortless everyday practices.[97] The representation of adulthood markers during the act rhetorically validates the new thematic notion, before a sense of mastery is achieved. "Permitting the mind to know the one by virtue of the other—is the talent of the master," Campbell argues.[98]

The first step, *The World Denied*, confronts the problem to accept that practical difficulties and responsibilities are inevitable parts of a reasonably happy life and that, with all its faults, this "world is nirvana."[99] When held accountable, the mother in *Room* first refuses to accept this and attempts suicide,[100] while the man in *The Road* temporarily abandons his belief in humanity and care in an act of vengeance. When the boy (Kody Smit-McPhee) reacts, the man cries out that he is "scared," for he is the one "who has to worry about everything."[101]

To convince of the new approach's relevance and application in the social world, a demonstration of practical "human success" is needed.[102] For that purpose, the next step, *Flight and Rescue*, emphasizes the pragmatic balance between the thematic markers in homelike environments. In *Room*, the mother struggles to trust in others, distantly observing and learning from her own mother and stepfather, as they teach the boy to cooperate, interact, show care, and follow social conventions. On the screen, Robbie's and Cecilia's "cheap and shabby" home in *Atonement* displays the symptomatic and necessary compromises that adult reality often requires.

In the *Return to the Normal World*, the new thematic notion is confronted with society and a "return blow of reasonable queries, hard resentment, and good people at a loss to comprehend."[103] Yet, with their solid self-esteem, the characters demonstrate that they accept both their own and others' faults, embracing that although "you can't change who people are," you can still "love them," as Louisa realizes in *Me Before You*.[104] This constitutes the last lesson regarding the complexity of the world that runs through the narratives. Even *Gone Girl*, an ironic tale about adulthood, matches this rhetorical pattern and adds an amplifying twist, as Amy and Nick finally accept that they make a pair of "a petty, selfish, manipulative, disciplined psycho bitch" and "an average, lazy, boring, cowardly, *woman-fearing* man."[105] The films and novels further accentuate the ultimate responsibility to a greater whole that being an adult entails and show how social adaptation and cooperation are tied to recognition and intimacy. Hence, Amy explains that she and Nick must bend to the fact that their efforts and reunion is "the story the world needs right now."[106] *Shutter Island*'s Teddy returns from his make-believe world and acknowledges how his weaknesses contributed to his wife's and children's death, which finally enables him to choose to live or die rather than to exist in a liminal zone.

In the final steps, *Master of Two Worlds* and *Freedom to Live*, *Gone Girl*'s couple performs intimate and cooperative acts of mind reading on the screen, despite the tensions between them. "That's how well you know me: You know me in your marrow," Amy concludes.[107] As *The Da Vinci Code*'s protagonists enter the chapel of Rosslyn to find the mystery's final key, they sense each other's thoughts and intentions with the same closeness and display of mastery. In *Atonement*, the internal author and narrator describes how she has chosen between two alternative endings, guided by inner values and beliefs. For her, it is "kindness,"[108] a word that implies both intimacy and care, that ultimately leads to happiness and makes it possible for her to conjure the dead lovers back to life—in the novel to a community of home, friends, and family; in the film to a state of secluded togetherness in harmony with nature. This belief in mankind and a balanced set of adulthood markers in the context of normal life, as keys to the elixir of happiness and *Freedom to Live*, is a conclusion that the narratives share. Near the end, Whatney describes how he embraces adult life's lull of responsibilities, through his experiences on Mars: "I figured out how to survive, at least for a while, and I got used to how things worked. My terrifying struggle to stay alive became somehow routine. Get up in the morning, eat breakfast, tend my crops, fix broken stuff, eat lunch, answer e-mail, watch TV, eat dinner, go to bed."[109] Nevertheless, it is the concluding thought that "every human being has a basic instinct to help out"[110] that reveals he can build a new life on Earth in a somewhat happy state of adulthood.

## Conclusion

As Thomas Leitch makes abundantly clear, one of the strengths of adaptation studies is that it allows for a number of alternative definitions and a productive variation of approaches.[111] The urtext concept abides by this principle and opens up for the examination of the web of nodes formed by a limited number of versions of a specific title, as well as for more inclusive studies. It can be used for explorations of genres, themes, and media conventions. For instance, Hogan has used an urtext approach to identify heroic, sacrificial, and love stories as the three dominant "universal narrative prototypes" with subgenres, all of which share the purpose to teach happiness.[112]

In this chapter, I have made use of Campbell's depiction of the monomyth urtext to demonstrate its usefulness as a rhetorical structure for stories' thematic arguments, beyond its common use as a model for prescribed character arcs and sequences of events. The diverse films and novels in my sample form a coherent web with clear thematic patterns and variations that add to the already existing map of the monomyth. As a result of the thematic approach, a general rhetorical structure reveals itself in the hero's journey, by which stories can be easily translated to rhetorical journeys through any chosen theme. Although adulthood has been the focus in my study, markers of gender, class, emotions, or any other theme can be applied to explore the thematic line of reasoning in fictional works.

What, then, is the core of the specific argument about adulthood that the ten pairs of adaptations reveal? First, the films and novels represent adulthood as a state that does not give much room for self-esteem, cooperation, care, commitment, or plans for the future in everyday life. The intuitive notion of adulthood instead revolves around work, home, and family; rationality; and simplifying demarcations to avoid unnecessary complexities. Yet, the call is made for a more complex worldview, associated with a richer emotional and cooperative social mindset. In the second act, the novels and films demonstrate that the markers of adulthood are interdependent. Self-esteem, self-awareness, and a consciousness of the social world's complexities are presented as key factors to implement this understanding, while rigid obedience to conventions obstructs. In the third act, the stories tell us that substantive qualities, especially beliefs, a solid worldview, and emotional balance, are necessary to develop the sense of adulthood further. With those elements in place, a fully developed care for others, a sense of intimacy, and autonomy of mind within a frame of tolerant conventions can be fully integrated. Finally, in the fourth act, the stories demonstrate the importance of compromise and acceptance of everyday life's imperfections for a somewhat happy state of adulthood to be accomplished.

Each film and novel naturally displays variations that add nuances to the adulthood urtext, both as individual texts and as adaptation pairs. There are also certain disparities between the films and novels. In general, the films afford more weight to social markers, home, and family, perhaps as a consequence of the possibility to better represent the characters visually. On the other hand, the novels accentuate the need for awareness of complexities and the substantive role of values and beliefs. Yet, both novels and films produce equally multifaceted rhetorical representations of what a gratifying adult life generally demands.

That the selected novel and film corpora reveal great similarities confirms the validity of the monomyth as a rhetorical model for thematic adaptation studies, without any partiality toward novels or films. Although I argue that the monomyth particularly invites adulthood as a theme, the rhetorical approach is universally applicable to other themes and their respective markers to describe how they are represented in multiple variations of a story or an urtext. This naturally invites further questions regarding the use of novels' and films' expressions to articulate thematic functions and dimensions in relation to different rhetorical steps, especially in a media landscape that is getting more and more niche and theme oriented. The model offers opportunities for structured corpora analyses of thematic arguments in particular genres and case studies, for instance, to explain the failure or success of an adaptation to connect with specific audience groups. For screenwriters who adapt novels, it can serve as an analytical tool to catch the spirit or communicated meaning of a text.

Regardless of approach, the final lesson is that novels and films are generally equally complex in their thematic investigations. When a film and novel are regarded as adaptations of a mutual urtext, they tend to complete each other, as if the characters and their narratives have more to say about the main theme, and, like real people, they change their viewpoint slightly when they migrate to a new story environment. Therefore, when they are examined, the decisive factor should not be the prejudice for a specific version or form but what the versions actually communicate, what alternative perspectives they present, and how we relate to that. In terms of adulthood, the various versions discussed here reveal "a vision transcending the scope of normal human destiny and amounting to a glimpse of . . . life as a whole."[113]

## Notes


1. Patrick Colm Hogan, *Affective Narratology: The Emotional Structure of Stories* (Lincoln: University of Nebraska Press, 2011), 182, 211.

2. Dennis Dutton, *The Art Instinct* (Oxford: Oxford University Press, 2009), 110; Hogan, *Affective Narratology*; and Blakey Vermeule, *Why Do We Care about Literary Characters?* (Baltimore: Johns Hopkins University Press, 2010).

3. Linda Hutcheon, *A Theory of Adaptation*, 2nd ed. (Oxon: Routledge, 2013), xvii.

4. Lars Elleström, "Adaptation and Intertextuality," in *The Oxford Handbook of Adaptation Studies*, ed. Thomas Leitch (New York: Oxford University Press, 2017), 524.

5. Sarah Cardwell, *Adaptation Revisited: Television and the Classic Novel* (Manchester: Manchester University Press, 2002), 26.

6. Cathlena Martin, "Charlotte's Website: Media Transformation and the Intertextual Web of Children's Culture," in *Adaptation in Contemporary Culture: Textual Infidelities*, ed. Rachel Carroll (London: Continuum, 2009), 90.

7. See Jerome Bruner, "Life as Narrative," *Social Research: An International Quarterly*, 71, no. 3 (2004); and Dan McAdams, *The Redemptive Self: Stories Americans Live By* (Oxford: Oxford University Press, 2006).

8. Joseph Campbell, *The Hero with a Thousand Faces*, 3rd ed. (Novato: New World Library, [1949] 2008).

9. Ibid., 12.

10. Ibid., 29.

11. I adapt Thompson's four-act structure labels to underscore the rhetoric progression through an *Introduction, Complication, Development*, and *Conclusion*. See Kristin Thompson, *Storytelling in the New Hollywood: Understanding Classical Narrative Technique* (Cambridge, MA: Harvard University Press, 1999).

12. Campbell's text has served as a prototype for many writing manuals, for example, Christopher Vogler, *The Writer's Journey: Mythic Structure for Writers*, 3rd ed. (Studio City, CA: Michael Wiese Productions, 2007), and Craig Batty, *Movies That Move Us* (Basingstoke: Palgrave Macmillan, 2011), and for feature films and novels alike, most famously George Lucas's *Star Wars* saga.

13. See, for example, Jeffrey Jensen Arnett, *Emerging Adulthood: The Winding Road from the Late Teens through the Twenties* (Oxford: Oxford University Press, 2004); Lewis R. Aiken, *Human Development in Adulthood* (New York: Plenum Press, 1998); Harry Blatterer, *Coming of Age in Times of Uncertainty* (New York: Berghahn Books, 2007); Varda Konstam, *Emerging and Young Adulthood* (Heidelberg: Springer, 2015); Erik H. Erikson, *Identity and the Life Cycle* (New York: Norton, [1959] 1980); and James A. Côté, *Arrested Adulthood* (New York: New York University Press, 2000).

14. The qualities that Campbell mentions as significant for adults (e.g., stability, rationality, independence, responsibility, cooperation, and authority) form an alternative set of useful markers. In *Visions of Virtue in Popular Film* (Boulder, CO: Westview Press, 1999), Joseph H. Kupfer makes good use of Aristotle's list of virtues.

15. Gérard Genette, *Narrative Discourse: An Essay in Method*, trans. Jane E. Lewin (New York: Cornell University Press, 1980); Hogan, *Affective Narratology*; and Brian McFarlane, *Novel to Film* (Oxford: Clarendon Press, 1996).

16. James Phelan, *Reading People, Reading Plots: Character, Progression, and the Interpretation of Narrative* (Chicago: University of Chicago Press, 1989).

17. Rick Altman, *A Theory of Narrative* (New York: Columbia University Press, 2008).

18. As Matthias Brütsch demonstrates in "The Three-Act Structure: Myth or Magical Formula?" (*Journal of Screenwriting* 6, no. 3 [2015]: 301–26), intersubjectivity always plays a part when plot steps are identified.

19. Market success, audience impact and scores, awards, and professional reviews are a few of the many ways to approach the problem. For some titles, initially considered, the novel met a certain quality criterion, but the film did not, or the other way around. In the end, a subjective aggregate had to be made of all these criteria.

20. Ian McEwan, *Atonement* (London: Jonathan Cape, 2001).

21. E. L. James, *Fifty Shades of Grey* (New York: Vintage Books, 2012).

22. Gillian Flynn, *Gone Girl* (London: Phoenix, 2012).

23. Jojo Moyes, *Me Before You* (London: Penguin, 2012).

24. Emma Donoghue, *Room* (London: Picador, 2010).

25. Dennis Lehane, *Shutter Island* (London: Bantam Press, 2003).

26. Dan Brown, *The Da Vinci Code* (London: Corgi Books, 2003).

27. Andy Weir, *The Martian* (London: Random House, 2014).

28. Cormac McCarthy, *The Road* (New York: Random House, 2007).

29. Walter Kirn, *Up in the Air* (New York: Doubleday, 2001).

30. Campbell, *Hero with a Thousand Faces*, 43.

31. Moyes, *Me Before You*, 30, and film minute 8.40.

32. James, *Fifty Shades of Grey*, 11, and film minute 7.10.

33. Ibid., 24.

34. Film minute 3.40.

35. Kirn, *Up in the Air*, 27.

36. Film minute 16.50.

37. Film minute 22.30.

38. Weir, *The Martian*, 73.

39. Flynn, *Gone Girl*, 81.

40. Brown, *Da Vinci Code*, 144.

41. McEwan, *Atonement*, 93.

42. Film minute 29.50.

43. Campbell, *Hero with a Thousand Faces*, 64–73.

44. Film minute 26.50.

45. Lehane, *Shutter Island*, 107.

46. Campbell, *Hero with a Thousand Faces*, 77.

47. Film minute 42.00.

48. Campbell, *Hero with a Thousand Faces*, 81.

49. Vogler, *Writer's Journey*, 6, 135.

50. Moyes, *Me Before You*, 178.

51. Donoghue, *Room*, 131.

52. Film minute 41.30.

53. James, *Fifty Shades of Grey*, 150.

54. Film minute 29.56–31.58.

55. Flynn, *Gone Girl*, 135.

56. Moyes, *Me Before You*, 141.

57. Ibid., 205.

58. McCarthy, *The Road*, 80.

59. Brown, *Da Vinci Code*, 174.

60. Campbell, *Hero with a Thousand Faces*, 97.

61. Ibid., 95.

62. James, *Fifty Shades of Grey*, 221.

63. Weir, *The Martian*, 183.

64. Ibid.

65. Film minute 45.50.

66. Campbell, *Hero with a Thousand Faces*, 101.

67. Moyes, *Me Before You*, 213.

68. Campbell, *The Hero with a Thousand Faces*, 101.

69. Lehane, *Shutter Island*, 216.

70. Donoghue, *Room*, 193.

71. McCarthy, *The Road*, 139.

72. Film minute 59.00.

73. Ibid., 69.24.

74. McEwan, *Atonement*, 187.

75. Brown, *Da Vinci Code*, 300.

76. Campbell, *Hero with a Thousand Faces*, 110.

77. Film minute 1.25.50.

78. Kirn, *Up in the Air*, 172.

79. Film minute 1.12.16.

80. Film minute 1.10.20.

81. Film minute 1.02.15–1.07.35.

82. Campbell, *Hero with a Thousand Faces*, 137.

83. Moyes, *Me Before You*, 349.

84. Film minute 1.14.20.

85. Film minute 85.00.

86. Campbell, *Hero with a Thousand Faces*, 133.

87. Brown, *Da Vinci Code*, 357.

88. James, *Fifty Shades of Grey*, 355.

89. Brown, *Da Vinci Code*, 390.

90. Film minute 1.14.30–1.15.

91. Campbell, *Hero with a Thousand Faces*, 150.

92. McCarthy, *The Road*, 224.

93. Lehane, *Shutter Island*, 293.

94. Film minute 74.47.

95. Kristin Thompson, *Storytelling in the New Hollywood: Understanding Classical Narrative Technique* (Cambridge, MA: Harvard University Press, 1999), 29.

96. Campbell, *Hero with a Thousand Faces*, 170.

97. Ibid., 167.

98. Ibid., 196.

99. Ibid., 141.

100. Donoghue, *Room*, 193, and film minute 1.33.

101. Ibid., 277, and film minute 1.22.47.

102. Campbell, *Hero with a Thousand Faces*, 178.

103. Ibid., 186.

104. Film minute 1.37.13.

105. Flynn, *Gone Girl*, 441.

106. Ibid., 443.

107. Film minute 2.12.12.

108. McEwan, *Atonement*, 371, and film minute 1.48.22.

109. Weir, *The Martian*, 341.

110. Ibid., 369.

111. Thomas Leitch, "Adaptation and Intertextuality, or, What Isn't an Adaptation, and What Does It Matter?" in *A Companion to Literature, Film, and Adaptation*, ed. Deborah Cartmell (Oxford: Wiley-Blackwell, 2012), 87–104.

112. Hogan, *Affective Narratology*, 125.

113. Campbell, *Hero with a Thousand Faces*, 201.

# 10

# Experimental Music as a New Frontier of Adaptation Studies

## Thomas Britt

MUSIC IS AN INFINITE subject. While many listeners and scholars have tried to quantify the highest possible number of melodic arrangements and combinations, there is no proven limit to how many compositions can be written or generated. To fully calculate the possibilities transcends human intelligence and the time and energy needed to reach such a conclusion. In *The Shattered Self: The End of Natural Evolution*, Pierre Baldi proposes a materialist's prognostication about human intelligence as a whole: "One day it might be viewed as a historically interesting, albeit peripheral, special case of machine intelligence."[1]

An intriguing phrase about temporality to emerge within recent musical history is *Years Past Matter*,[2] the title of experimental black metal band Krallice's fourth studio album.[3] One way to read the phrase is to envision a future state of total annihilation, of form lost, becoming void. But the other way to read the title is to appeal to individual and collective memories, to remember the past as being consequential. Indeed, the alternately melodic and dissonant sound of *Years Past Matter* is akin to being pulled toward past and future at once, with the attendant physical and psychic tension one might expect. This tension, between what was and what will (not) be, is similarly present in Jacques Derrida's "hauntology," popularly remembered as "the first major trend in critical theory to have flourished online"[4] and associated with another intriguing phrase: "lost futures."[5]

Another theoretical context for this arrangement of past and present is John Ellis's description of "the aim of adaptation" in his introduction to "The Literary Adaptation" from *Screen*:

> The adaptation trades upon the memory of the novel, a memory that can derive from actual reading, or, as is more likely with a classic of literature, a generally circulated cultural memory. The adaptation consumes this memory, aiming to efface it with the presence of its own images.[6]

The process Ellis describes here envisions something (a source text) becoming nothing (as it is erased and replaced) becoming something new (the adaptation of the source text).

Academic work on musical adaptations is rather marginalized within adaptation studies, which is mainly restricted to film adaptations of literature or other narrative sources. A key difference between those two spheres is that evaluation of musical adaptations is not as focused on fidelity, which guides much of the discussion in film adaptations of literature/narrative. Musical adaptations and the study thereof are diverse in content and theoretical approach, with some of the diversifying factors being conditions of verbal versus nonverbal musical forms and the transformative possibilities of turning written texts or spoken narratives into musical sounds (and vice versa).[7]

Regarding popular music forms, which often veer from classical sources and texts into more contemporary source materials, the means of production have determined their aesthetics, ethics, and evaluative standards. Experiments with tape loops, vinyl records, and other analog techniques have progressed into various digital technologies to further establish and evolve what is known today as sampling in popular music genres into the mash-up approach to transforming existing material and to go beyond to include the latest incarnation, the type beat. Sampling, as it exists within the presently dominant genres of hip-hop, rap, and dance music, is to some degree influenced by the longtime practice of quoting existing music within rock and roll compositions. These areas are likewise distinct from film adaptations of literature/narrative, in part because of their comparatively straightforward involvement of the conditions for "fair use," including a measurable portion of the existing work being used and especially the measurable degree of transformation present within the act of adapting.

In "Amen to That: Sampling and Adapting the Past," Steve Collins describes the "constant principle" of a range of standard sampling techniques thus: "An existing work is appropriated and adapted to fit the needs of the secondary creator."[8] He goes on to state that "sample-based music—as an exercise in adaptation—encourages a Foucauldian questioning of the composer's authority over their musical texts."[9] While those conditions certainly relate to the "mash-up" genre that Collins considers, my aim is to investigate musical works that have a different relationship to source texts and the authority therein.

Of the examples considered here, the Vaporwave genre and resultant/related musical and nonmusical memes most align with traditional approaches to sampling and mash-ups, albeit in a highly ironic/anarchic context. The other works examined in this chapter involve a sole music creator creating a source and then adapting it for another purpose of their own (William Basinski, Phil Elverum of Mount Eerie, and Sam Kidel), musical creators imagining a source that is unavailable to them (Toyomu and the Fiery Furnaces), or a sole musical creator adapting nonmusical sources into music (PJ Harvey). The variety within these case studies illustrates the fluidity beyond fidelity discourse that is a hallmark of musical adaptations, both in practice and in study. Basinski and Elverum transform music over which they already had compositional and emotional authority to pose questions about destruction and death (perhaps the ultimate loss of authority). Adaptations by Toyomu and the Fiery Furnaces introduce the possibility that imagination itself is a sufficient seed for adaptation, for creating something from nothing. Harvey uses musical adaptation to turn observation into satirical social realism.

My goal is to establish experimental music adaptation as a new frontier in adaptation studies because the creation and analysis of such works introduce some innovative approaches to production and distribution. These works do not depend on creating or quoting visual information or traditional narrative, nor are they evaluated with regard to fidelity to existing narratives. The aim is to point out the way the adapting artists are able to adventurously explore aural and temporal qualities and use networked communication to share their adaptations. My methodology alternately contextualizes this music within network theory, graph theory, adaptation theory, and philosophies of authenticity and information transfer. I identify a number of albums (some only available as digital information) that share some common adaptation characteristics and could probably be depicted as vertices in a single graph of contemporary experimental music. However, as each of these individual works involves adaptation, I examine them separately while acknowledging that they share qualitative properties. As I investigate each example, I focus on the direction of the adaptation, the distance-time of the adaptation, and the degree of the adaptation. Alongside this track of dimensional information is another track of theoretical or philosophical inquiry, a framework informed by additional definitions of adaptation, appropriation, and reproduction that I will now introduce.

The first is an engagement with adaptation similar to Thomas Leitch's approach in *Film Adaptation and Its Discontents*—that is to say, embracing the intertextuality of adaptation and thinking through the area of production, consumption, and critique that is "between adaptation and allusion,"[10] particularly in the many ways the adaptations considered here make "adjustments" to their source material. The second is an understanding of appropriation in the manner defined by Julie Sanders as being distinct from adaptation, "a more decisive journey away from the informing source into a wholly new cultural product and domain."[11] The third is from Linda Hutcheon, in the "Final Questions" chapter of *A Theory of Adaptation*. Hutcheon theorizes that "adaptation is how stories evolve and mutate to fit new times and different places. . . . Temporal precedence does not mean anything more than temporal priority."[12] The fourth underpinning for this chapter is Walter Benjamin, whose influences[13] and ideas about materiality and meaning and reflections on the "aura" of works of art are particularly useful for exploring the musical works considered here. Though the specific meaning of Benjamin's concept of "aura" has been contested, Benjamin scholars coalesce around some basic conditions: "the aura is the manifestation of authenticity as a mode of separation,"[14] and "the aura includes a sensory experience of distance between the [spectator] and the work of art."[15]

William Basinski's *The Disintegration Loops*[16] is an artwork in which the "aura" and other Benjamin themes are freshly realized. Beginning in the 1980s, Basinski created a number of sound loops (of various sources) on magnetic tape. He hung them from a tree branch by his desk and left them there. These constitute the original material that he would later adapt. Within graph theory, it's useful to think of these loops as tactile self-loops. In the summer of 2001, Basinski made digitally captured copies of the original works, and the process recorded the physical deterioration of the magnetic tapes. This digital capture is the process and product of the adaptation of the original material. In essence, the original aura was destroyed, but the process of that destruction was preserved by the sounds of dying loops, archived in digital perfection, the birth of a new aura. The act of preserving, like a rescue mission, here fails to preserve in a lossless way but instead enables new digital creation from physical decay. The original loops' experience toward vanishment is fascinating, as each self-loop varies in character, never perfectly replicating, each one functioning as a distinct adaptation of a vanishing original. The overstated part of *The Disintegration Loops*'s legacy clouds that primary contribution to the history of experimental music. Basinski publicly positioned his tape transfer effort of 2001 as a 9/11 event. He has spoken about his activity and state of mind on September 10, 2001, "listening to the work, swimming in it . . . knowing I had this work I didn't know what to do with."[17] The next day, with a view of the attack from his "Brooklyn rooftop," he "was determined that it was going to be an elegy."[18] To concretize that form of expression, he joined the sound of the dying tapes with one other work, which was just over an hour's worth of video footage he shot of fresh smoke rising from the ruins of the World Trade Center site. Basinski chose to adapt his private moment of observation into a mediated event, appropriating his individually remembered musical past into the collective experience and memory of the present age of terror.

*The Disintegration Loops* was released beginning in 2002 as an album on CD and vinyl formats (in multiple installments) with an accompanying DVD of 9/11 aftermath footage set to the music, available to consumers in deluxe editions for hundreds of dollars. The direction of adaptation for *Disintegration Loops* is fairly straightforward, from analog source to digital source to analog and digital distribution formats, though the eventual dual directedness into musical and visual forms negatively affects the unity of form and content.

More complex is the distance-time of *The Disintegration Loops*. For years, this music hung from a tree branch as a discrete object available only to its creator and unaffected by technological advances of digital recording and advances in networked communication. Then, across a few months of a single year, the loops encountered innovations, which both destroyed them and gave them new life. But their reception (their temporal priority) was settled when Basinski imposed the 9/11 narrative on them, essential freezing the works to that moment in time, though deluxe editions would develop in the decade afterward and YouTube plays of his rooftop video accumulate to the present day. Each sale and hit does not count as a new incident contributing to the degree of each node in this process of adaptation, but one could receive the commercially available *Disintegration Loops* as having mutated into a different sort of self-loop—a work that experienced unlikely transformations only to become so fixed to historical conditions that its other formal significances are variants that did not survive under the weight of an appropriated historical trauma.

A few years after the original release of the *Disintegration Loops*, a microgenre called Vaporwave emerged. Vaporwave is a distinct form/area of exploration from *Disintegration Loops*, but they are related insofar as Basinski and Vaporwave are explicitly and implicitly concerned with the 9/11 terror attack and its implications. The source material for Vaporwave releases includes passé adult contemporary music of the 1980s and 1990s, appropriated and repurposed toward a haunted aesthetic and obfuscated behind fake artist names, foreign languages, and anachronistic artwork merging structures and statues of ruin with fluorescent colors and icons of the early PC era. These aural and visual objects give direction to an adaptation of the collective memory/"lost futures" of consumers from the 1980s and 1990s, specifically those who lived through and/ or long for the years of Japan's economic peaks and subsequent lost decade, as well as the last decades in which the United States' World Trade Center would be seen as a "monument to capitalism," before being destroyed and reborn as "a monument not to capitalism but to what happened on September 11th."[19] Perhaps the most popular release of the microgenre was *Floral Shoppe*[20] by American electronic musician Vektroid, released under a pseudonym.

Much can be said about how quickly Vaporwave rose and fell in conjunction with the Internet meme era, a new era of irony, of actively signifying ownership of cultural information. Such ownership might be seen as another digital evolution of the contested "composer's authority"[21] involved in sampling existing musical sources. A comprehensive *Esquire* article, "How Vaporwave Was Created Then Destroyed by the Internet," explained the microgenre's brief history:

> Vaporwave arose in reaction to huge economic and social forces
> that are still very much a part of our lives: globalization, runaway
> consumerism, and manufactured nostalgia chief among them.
> There is no other kind of music that explicitly concerns itself
> with these aspects of our zeitgeist.[22]

That outline of forces is a good starting point for explaining what historically informs the source material of Vaporwave and what its proponents adapt those sources into. It is worth pointing out, however, that other popular genres of music that preceded Vaporwave, namely, much hip-hop and rap, are equally explicitly concerned with the present globalist capitalist realities, but from a different perspective.

In *Babbling Corpse: Vaporwave and the Commodification of Ghosts*, Grafton Tanner describes Vaporwave's medium-specific existence for the Internet meme era: "Vaporwave is the music of 'non-times' and 'non-places' because it is skeptical of what consumer culture has done to time and space."[23] In this way, Basinski's approach to adaptation—collapsing the dimensions of his work and affixing the work to ruin—is not so different in its effects from the economic and cultural decline on which Vaporwave apparently asks us to reflect. The distance-time of the adaptation process for Vaporwave involves a period of a couple decades, or roughly a single generation.

There's something culturally revealing about what the next networked generation does with Vaporwave's source materials, which is to adapt them with new degrees of ideological appropriation, producing provocative and racist Internet memes. The most acute example is the Moon Man meme (and associated body of media) appropriating a character (Mac Tonight) from a McDonald's advertising campaign originally active from the mid-1980s to early 1990s. For the half decade preceding Vaporwave's emergence, online users and meme creators on *YTMND*, an early and influential online meme community, had used the visual object of the Mac Tonight character in conjunction with aural information, such as existing hip-hop and rap music and spoken words/ lyrics generated with text-to-speech utilities (especially one made available by AT&T). Then Vaporwave artist Saint Pepsi combined video footage of a Mac Tonight television advertisement with the song "Enjoy Yourself,"[24] a release that amassed millions of views on YouTube and arguably sustained the appropriated/ adapted imagery for its next rebirth.

The Mac Tonight advertising campaign, effaced and adapted, soon became a cultural force that *Salon* labeled "the alt-right's racist rap sensation."[25] That sensation, popularly known as Moon Man, appropriated rap lyrics and/or spoke original propaganda in a uniformly machine-man vocal style, delivering nationalistic, racist, and anti-Semitic messages. The viral success of "Moon

Man" videos, fueled by image boards and video sharing websites, is evidence that the same cultural detritus mined for the reflection of a collective consumerist memory can adapt to be weaponized. For citizens concerned about cultural rot, the evolution of "Moon Man" media might serve as evidence that the non-time and non-place[26] of the Internet has sapped humans of their capacity for ethics.

Beyond the instances of such ideological hijacking, the core theme of post-human alienation remains a fruitful subject matter for experimental music adaptation. Phil Elverum of Mount Eerie, one of American independent music's foremost folk poets, who writes on subjects of death and decay in the natural world, commented on adaptations that concretize and aestheticize a post-human void by mechanically adapting his own earlier material to create the album *Pre-Human Ideas*.[27] Elverum promoted the album thus:

> This is an unusual release for Mount Eerie for many reasons. Here are some of them:
>
> > -overtly recorded on a computer (as opposed to the customary extreme 100% analog style)
> > -pretty weird human portrait on the cover (as opposed to a non-personal place-focus)
> > -totally synthetic voices, digitally garbled (as opposed to legible and literate human tones)
> > -no new songs (as opposed to the perception that Mount Eerie is "prolific")
> > -there is no such thing as a difference between 'nature' and 'not nature', it is all wild.[28]

By promoting his act of adaptation with a series of oppositions contrasting the present work with his past work, Elverum positions the listener to focus on both the process and the product while listening. The direction of adaptation for *Pre-Human Ideas* is similar to *Disintegration Loops* insofar as analog sources become digital products and are then released in a variety of analog and digital formats. A key difference between the two works is that Basinski's tape transfer process digitally preserved the dying sounds of his decaying earlier texts, whereas Elverum *replicates* (ironically-technologically "upgrades") his earlier texts using a markedly different set of production tools than that with which they were originally realized.

There are many ironies involved in Elverum's choice to call this set of recordings *Pre-Human Ideas*. Chiefly, listeners not already aware of the pre-existing primary analog versions of the songs might perceive these recordings as initial "demo" versions intended as demonstrative testing ground for new musical ideas. In this way, *Pre-Human Ideas* shuffles the distance-time involved in the act of adapting and creates the (functionally illogical but fascinating to consider) form of an ex post facto demo recording. While critics' assessments of *Pre-Human Ideas* both positive and negative have pointed out the true demonstration purpose of these recordings, which was to provide instruction to musicians as they learned these songs to tour live with Elverum, *Pre-Human Ideas* transcends that purpose in the way it invites inquiry into philosophies of composition, recording, performance, and adaptation that run throughout Elverum's discography.[29]

In a 2014 interview, Elverum says, "The recordings always come first. I almost always compose songs as I'm recording them, so they don't really exist before they're recorded. . . . So live versions are always secondary pale echoes of the 'real' recorded versions."[30] Later in the interview, he says of adapting his own music, "Context is everything. Also, songs are evolving living things. Also nothing is true or real. It's all just many layers of strangers misunderstanding each other's poetry."[31] This later statement, in particular, aligns with a worldview present in Mount Eerie song "Ancient Questions" from the black metal–inspired album *Wind's Poem*: "Nothing means nothing / Everything is fleeting / Don't get used to it / I say, look upon the ruins."[32]

*Pre-Human Ideas*, promoted by the artist as a work opposed to his standard output, creates a number of questions by showcasing the contradictions and oppositions that also appear in his statements to Mescher. If the live performances are "secondary pale echoes" of the recorded versions, then what to make of new recordings that intentionally make use of synthetic and digital techniques and aesthetics? Are they later, tertiary echoes? Or do they offer a way to hear these songs from a new, primary starting point? If "nothing is true or real," then the evaluative standard of realness with which Elverum determines the primacy of the original recordings no longer matters. Perhaps the most interesting evolving factor related to *Pre-Human Ideas* is the reality that Elverum's preoccupation with death and decay has grown considerably more intense since the death of his wife and her memorialization in the diaristic Mount Eerie albums *A Crow Looked at Me*[33] and *Now Only*.[34] These are albums about what happens to the body after a person dies (it becomes dust and bone) and what happens to those who survive (their memory of the deceased revives what is now dust and bone). As an artist, Elverum's present concern is what it is for a human person to become nothing. Listening to *Pre-Human Ideas*, a distinctly nonhuman-sounding album that exhibits a machine-man's limited emotional expressiveness on subjects of decay and the natural world, one is haunted by the specter of digital technology that will outlive and exceed the human. It's the sound of a machine looking upon the ruins of humanity. Though *A Crow Looked at Me* and *Now Only* are works of and about mourning, they are also albums in which Elverum adapts the material and memory of his dead wife into a new form; he revives her in song.

In 2016, Sam Kidel synthesized the aesthetic qualities of both Basinski and Vaporwave with the machine-man ideas of Mount Eerie/Phil Elverum, creating the most directly adaptable and mutable work of art discussed here. *Disruptive Muzak*[35] is an album the artist created after exploring research and methods of the Muzak Corporation. In this recording, he places phone calls to the Department for Work and Pensions in the United Kingdom and, instead of speaking, plays his own ambient composition (perhaps his own variation of Muzak) to the humans, machines, and, indeed, hold music that he encounters on the phone. *Disruptive Muzak* is in part a recording that adapts an original Kidel composition into a series of dialogues between humans and music, evoking a sense of longing to connect. Yet, the initiation of the phone call is not necessarily synonymous with the initiation of communication. In addition, the communication therein is "one-sided" only if we consider words to be the only way to communicate on a phone. By making an album about the distance-time between two parties brought together through a communication network, and then using only music to respond to those answering with words, Kidel documents a series of brief encounters and then edits them into the "A" side of his album. As an adaptive technique, this form of compression preserves the aura of the original because the finished recording creates the network of alienation that is the project's form and content.

*Disruptive Muzak* is also complicated in an ethical sense by the way concepts of authority and consent have traditionally been involved with Muzak and critiques thereof. In short, consumers play their role by being controlled by advertisers and psychological manipulation, and not all consumers are aware of how they are being researched and being sold. The use of Muzak to guide mood, decisions, and concentration, among other effects, has long been debated within theories of consumption and related fields. But Kidel's approach to critiquing Muzak is to in some ways adapt his musical project into committing the same violations that critics of Muzak claim the form commits. So, rather than be forced to listen to hold music while trying to access a government department on which millions depend, Kidel subverts the operators' expectation for human cooperation in trying to do their job, which is to connect people. Kidel is acting as a serial disconnector, denying the kind of access the operators expect. They speak their greetings and questions and puzzlement into a void that only responds with ambient music. The effect is a reflection of the sort of alienation that Muzak is accused of contributing to, but none of the operators have consented to participate in an artistic reflection. Additionally, as a source for future adaptation, *Disruptive Muzak* opens up to infinite degrees because the "B" side of the album is a DIY version of Kidel's composition that allows any listener to perform their own *Disruptive Muzak* in the manner of the original but with a necessarily varying network of responders.

PJ Harvey is another musician who has recently communicated an artistic/political aim in part by manipulating, perhaps breaching, the implied rules of consent that guide the forms and texts she adapts into music. In a 2016 article titled "I Gave a Famous Rock Star a Windshield Tour of D.C.—and Didn't Know Who She Was," the writer Paul Schwartzman describes having been contacted by "a war photographer named Seamus Murphy" who wanted to know if he would "lead a tour of Washington's roughest neighborhoods for him and a woman he described as a 'musician/poet.'"[36] The musician/poet was PJ Harvey, who was working alongside Murphy researching Washington, DC, and some other global locations to write her next album. By enlisting Schwartzman to provide the tour of the "roughest" parts of DC and then documenting what he said while being vague about the degree to which his contribution would feature in the final product, Murphy and Harvey assumed the authority to use a journalist's spoken words as content for an art experiment, eventually released as a musical album, *The Hope Six Demolition Project*,[37] and related media. The recording process for the album was promoted as "an architectural installation, open to public view at Somerset House in London for four weeks only," that involved "displaying through one-way glass, PJ Harvey, her band, producers and engineers as a mutating, multi-dimensional sound sculpture."[38] Harvey called the installation *Recording in Progress* and said she intended "to operate as if we're an exhibition in a gallery. I hope visitors will be able to experience the flow and energy of the recording process."[39]

That Harvey invited spectators to observe but not to record the installation is an additional example of how her authority superseded the will of other contributors/participants as potential creators. Her control of various time-and-space-based dimensions of the project, including two visits with Schwartzman, the days spent *Recording in Progress*, and other related research and recording activities might be seen as an approach to creative research that is disproportionately intent on appropriating material from real life and adapting it without inviting those around her to fully participate. At what point this becomes exploitation is for critics, and especially the involved spectators, to decide. An instance of music's relationship to time, *Recording in Progress* reengages with a simple, sometimes neglected, fact of music history that John D. Barrow recounts in *The Artful Universe*:

> We must not forget that the first music was what we would now call "folk" music: music that had not been deliberately composed or written down. It was not made for study or appreciation in the modern fashion: you listened to it only in order to learn how to participate in its performance.[40]

The music video for "The Community of Hope," a single from *The Hope Six Demolition Project*, is many things, including a partial substantiation of Schwartzman's account and an audiovisual travelogue and reimagining of that account, including his actual dialogue. It is also a music video that provokes a viewer to see the images of city life in conjunction with Harvey's lyrics about her tour through "Washington's roughest neighborhoods," lyrics that liberally quote and adapt Schwartzman's commentary that he provided in the context of Murphy's request to show him those types of neighborhoods. In this way, "The Community of Hope" is a subversive adaptation of the quality of objectivity prized in journalism. Had Harvey and Murphy's request been more general, it is likely that Schwartzman would have provided a more diverse view of the place and its inhabitants.

Above all, the music video is acidly satirical, as the crescendo features an authentic gospel choir singing, "They're gonna put a Wal-Mart here" over and over. The Wal-Mart lyric is adapted from a Schwartzman quote and becomes a biting indictment of the kind of salvation a community like this is promised by its political and economic leaders. In real life, plans to build the Wal-Mart were eventually scrapped, a point with which Schwartzman glibly concludes his article but which, on a theoretical level, creates another variation of a lost future. By using such a vulnerable place and population to make a satirical point, Harvey and Murphy (who directed the video) could be rightly criticized as individuals who exploited the community for their own artistic purposes.

In fact, after the video/single's release, community members including former mayor Vincent Gray and "Leah Garrett at DC the nonprofit Community of Hope initiative" cited the narrowness of Harvey's portrayal.[41] In her defense, I would claim that to turn the words of an embedded journalist into a melody that can't soon be forgotten is a form of activism that doesn't allow the listener to forget the need for change in a community whose future is threatened by many forces. During the controversy surrounding her portrayal of Washington, DC, Harvey appeared on BBC One's *The Andrew Marr Show*. Harvey described spending time in DC, "talking to people, listening to what they had to say, and I just collected notes as a journalist might."[42]

The final examples I explore are occupied by the realm of imagined sources. In 2016, Kanye West released *The Life of Pablo*,[43] a popular, critically and commercially successful rap album with Vaporwave-influenced cover art, which rolled out online in multiple versions across various platforms during several months. Well into the album's digital release, West continued to adapt—to further mix and edit—his own songs. Def Jam, West's label, called this work a "continuous process,"[44] which illustrates West's apparent interest in redefining the form and function of an LP within the commercial music industry, a creative inquiry similarly undertaken in recent years by other popular artists such as Radiohead and Trent Reznor. But the adaptation of *The Life of Pablo* considered here was a version not engineered by West at all.

In Japan, where CD sales are incredibly strong compared to other international markets for commercial music, West's digital-only album was not available to listeners because TIDAL, the platform for the initial exclusive digital release, had not been launched. So, experimental musician Toyomu used whatever sources were available to him, including the lyrics resource Genius and the sample database Whosampled, to explore the source materials for *The Life of Pablo* and to create a version of it for himself and for Japan. This research, only possible through the networked communication that allows such resources and databases, enabled Toyomu to track down the source sample texts and original lyrical content of *The Life of Pablo*, despite not having heard the original album. The result was *Imagining the Life of Pablo*,[45] an adaptation of an evolving source the producer had never heard but had imagined through reading information about the raw material of his source album's form and contents.

The vocals on Toyomu's album are not rapped but rather created through a text-to-speech application that connects this work to the approaches of other machine-man vocals and albums about post-human alienation. *Imagining the Life of Pablo* might be the experimental music release that most fully represents the capability of an interconnected global network and information transfer as cultural translation. In 2016, Reuben Tasker of Genius interviewed Toyomu about his methods: "Do you find it strange that, even in 2016, music cannot reach everyone? Is this a problem?"[46] The artist responded, "I don't think that it's problem. The world is so wide. If we can recognize about everything, that situation is stranger than now, dystopia. I believe imagining is happened in obscurity."[47] This statement is a lucid description of the aura still available in art that one hopes to adapt, the sensory experience of distance, foregrounding imagination as a meaningful source beyond, at times preferable to, the material.

Another example of an imagined source comes from Matthew Friedberger, one of two permanent members of the American pop/rock band the Fiery Furnaces, the other being Matthew's sister Eleanor Friedberger. While both siblings now enjoy solo careers, with the Fiery Furnaces having been on hiatus since 2011, one of the group's early albums, *Rehearsing My Choir*,[48] is a unique example of experimental pop music that features at least one song, "We Wrote Letters Everyday," adapted from an imagined source.

Between 2003 and 2009, the Fiery Furnaces released nine full-length LPs, most of which were received as dense and at times impenetrable works because of the overstuffed musical compositions and complex narratives. Even within that framework, *Rehearsing My Choir* stands out because it is an album-length biography of, and performed by, the Friedberger siblings' octogenarian grandmother, Olga Sarantos. Written by Matthew, the album shuffles the timeline of the source text, which is the recounted life of his grandmother. In the artist's statement and press kit for the album, Matthew delineates its structure:

> Dear Listener, Tracks 3 and 4 take place in the 40's; tracks 5 and 6 in the 20's and 30's; track 7 in the later 50's; track 8 starts in the very early 40's; track 9 goes back and forth; track 10 takes place in the early 60's; the final track takes place in the early 90's. Track 2 takes place a few years ago; track 1 took place when it was recorded.[49]

The direction of this adaptation is an oral history, with some extant supporting evidence, being turned into a pop/rock opera or concept album of sorts. As is evident in his statement, one of the key creative decisions Matthew made when adapting the life story of his grandmother was to reorder the telling of events that she lived chronologically. The decades of real life that constitute the distance-time of the adaptation are compressed into just less than an hour of music. Frequently, the recollections include precise details such as names and addresses of the people involved, as well as dialogue spoken at the time of the events, which are impressively integrated into the compressed and reordered life story. Eleanor's voice is an additional expression of her grandmother's persona and contributes to an exciting oscillation between past events and present, dramatized memory of those events.

Four songs into this dizzying biographical adaptation comes "We Wrote Letters Everyday," a 1940s-set number in which Sarantos describes the time her husband "went out to the Pacific" while she "went back to Chicago to work on the railroad."[50] The grandmother speaks the verse of the song, but then she is joined in duet with her granddaughter Eleanor for the chorus, in which they both sing, "And we wrote letters every day / Which were later thrown away / And God knows what we wrote or what they said / But this is probably how they read."[51] What follows is Matthew's entirely instrumental (pianos and keyboards) adaptation of those letters. Later, passages of the song reveal the specific manner of the letters' decay, but the contrasting musical dialogue Matthew creates to replace the lost letters his grandmother and sister are singing about is an inventive use of music to replace lost language. Sarantos, the author of the source text, cannot remember what she wrote, and the letters are now gone, unavailable. So

Matthew, the author of her biography, imagines and revives the aura of those letters through the language of the piano, a medium-specific expression of musical biography. "We Wrote Letters Everyday" is an adaptation that revives a lost piece of family history, extending it to the next generation in an altered form.

The limits of music are unknown. The future of human contributions to music is uncertain. To speculate on the infinite possibilities of the form and our role in it is to encounter a reminder of how limited humans are in space, time, and natural computing power. In this chapter, I have examined artists undertaking experimental musical adaptations in ways that preserve, replicate, critique, repurpose, and/or imagine information from the past. The foreseeable future indicates that letters will continue to decay, memories will continue to fail, borders will continue to create barriers to communication, and some individuals, cities, and nations will continue to be stranded and abandoned to lost futures. However, as Toyomu points out, one essential quality that delays certain dystopia is the gift of imagination.

The act of adaptation revises/replaces that which came before and extends the life of former products and projects into the uncertain future. Skeptic Bergen Evans once wrote, "The sudden emergence of presumably extinct ideas reminds us . . . how near to darkness we really are."[52] Perhaps the most enlivening adaptations are made with the awareness that the apparent onset of darkness offers the opportunity to adapt, to transport cultural memory that is seen through a glass darkly, ever toward light.

Furthermore, current developments in experimental musical adaptation serve as a reminder to not take for granted the human role in the preservation and revival of source objects through acts of adaptation. Someday soon the machines might ask the humans not to participate. Consider the metal band Krallice, whose *Years Past Matter* informs the conceptual framework for this chapter. In 2017, the band's 2011 album *Diotima* became source material used by DADABOTS to train a bot to create new music.

DADABOTS, who "write programs to develop artificial artists," created its adaptation of *Diotima*, called *Coditany of Timeness*, as "part of a submission to NIPS 2017 Workshop for Machine Learning, Creativity and Design: 'Generating Black Metal and Math Rock,'" and specify on the Bandcamp web page for the release that "this album was generated with a recurrent neural network [SampleRNN (Soroush Mehri et al. 2017)] trained on raw audio from the album 'Diotima' by Krallice. All titles were generated by a Markov chain."[53]

In other words, DADABOTS trained artificial intelligence to adapt a previously existing metal album into a distinct new work of a similar style. In this instance, fidelity is one test of how successful the experiment was, and AI proves to be remarkably adept at mimicking/performing the musical style it learned in order to create an exciting and new arrangement of that music. This development, alongside similar efforts by AI to create visual art and narrative forms like screenplays, might see the human contribution to creative works erode against technologically superior standards of machine optimization and efficiency.

## Notes

1. Pierre Baldi, *The Shattered Self: The End of Natural Evolution* (Cambridge, MA: MIT Press, 2002), 113.

2. Krallice, *Years Past Matter*, self-released, 2012, CD.

3. Heavy metal music is a frequent source for adaptation from literary works. Notable examples include Mastodon's *Leviathan* (2004), an adaptation of Herman Melville's *Moby-Dick* (1851), and Earth's *Hex: Or Printing in the Infernal Method* (2005), an adaptation of Cormac McCarthy's *Blood Meridian* (1985).

4. Andrew Gallix, "Hauntology: A Not-So-New Critical Manifestation," *The Guardian*, June 17, 2011, https://www.theguardian.com/books/booksblog/2011/jun/17/hauntology-critical.

5. Mark Fisher, *Ghosts of My Life: Writings on Depression, Hauntology and Lost Futures* (Hants, UK: Zero Books, 2014).

6. John Ellis, "The Literary Adaptation," *Screen* 23, no. 1 (1982): 3, https://doi.org/10.1093/screen/23.1.3.

7. Two worthwhile examples of academic writing on these subjects within musical adaptations are Sammie Ann Wicks, "Music, Meaning, and the Adaptation of Literature," *Literature in Performance* 2, no. 1 (1981): 89–97; and Michael J. Meyer, ed., *Literature and Musical Adaptation* (Amsterdam: Rodopi, 2002).

8. Steve Collins, "Amen to That: Sampling and Adapting the Past," *M/C Journal* 10, no. 2 (2007), http://journal.media-culture.org.au/0705/09-collins.php.

9. Ibid.

10. Thomas Leitch, *Film Adaptation and Its Discontents* (Baltimore: Johns Hopkins University Press, 2007), 93.

11. Julie Sanders, *Adaptation and Appropriation* (New York: Routledge, 2006), 26.

12. Linda Hutcheon, *A Theory of Adaptation*, 2nd ed. (Oxon: Routledge, 2013), 177.

13. In *Walter Benjamin: The Colour of Experience* (New York: Routledge, 1998), 90, Howard Caygill discusses Alois Riegl's influence on Benjamin, specifically the distinction between "insights into the materiality of individual works of art and the metaphysics of the artistic will." Similarly, Erwin Panofsky in "The Concept of Artistic Volition" (trans. Kenneth J. Northcott and Joel Snyder, *Critical Inquiry* 8, no. 1 [1981], 19), describes Riegl's contribution to art history thus: "Rather than constantly emphasizing factors which determine the work of art—the character of raw materials, technique, intention, historical conditions—he introduced a concept which was to denote the sum or unity of the creative forces—forces both of form and content—which organized the work from within. This concept was 'artistic volition.'"

14. Gerard Watts, "Walter Benjamin's Definition of the Aura," *Digital Arts and Humanities (DAH) Structured PhD* (Royal Irish Academy), 2012, http://dahphd.ie/gerardwatts/author/gerardwatts/ (accessed May 2, 2018).

15. Andrew Robinson, "An A to Z of Theory: Walter Benjamin: Art, Aura and Authenticity," *Ceasefire*, June 14, 2013, https://ceasefiremagazine.co.uk/walter-benjamin-art-aura-authenticity/.

16. William Basinski, *The Disintegration Loops*, 2002–3, CD.

17. In Andy Battaglia, "From Destruction, a New Sound of Life," *Wall Street Journal*, September 10, 2012, https://www.wsj.com/articles/SB10000872396390444554704577643672237386032.

18. Ibid.

19. Richard Stengel, "Build a Monument to Capitalism," *Time*, January 11, 2002, http://content.time.com/time/nation/article/0,8599,192964,00.html.

20. Macintosh Plus, *Floral Shoppe*, Beer on the Rug, 2011, YouTube audio.

21. Collins, "Amen to That."

22. Scott Beauchamp, "How Vaporwave Was Created Then Destroyed by the Internet," *Esquire*, August 18, 2016, https://www.esquire.com/entertainment/music/a47793/what-happened-to-vaporwave/.

23. Grafton Tanner, *Babbling Corpse: Vaporwave and the Commodification of Ghosts* (Hants, UK: Zero Books, 2016), 39.

24. Saint Pepsi, "Enjoy Yourself," Illuminated Paths, 2013, YouTube audio.

25. Matthew Sheffield, "Meet Moon Man: The Alt-right's Racist Rap Sensation, Borrowed from 1980s McDonald's Ads," *Salon*, October 25, 2016, https://www .salon.com/2016/10/25/meet-moon-man-the-alt-rights-new-racist-rap-sensation -borrowed-from-1980s-mcdonalds-ads/.

26. Tanner, *Babbling Corpse.*

27. Mount Eerie, *Pre-Human Ideas*, P.W. Elverum & Sun, 2013, YouTube audio.

28. Ibid.

29. For positive reviews, see Jeremy D. Larson, "Mount Eerie: *Pre-Human Ideas*," *Pitchfork*, November 11, 2013, https://pitchfork.com/reviews/albums/18719 -mount-eerie-pre-human-ideas/. For negative reviews, see Alex Young, "Mount Eerie—*Pre-Human Ideas*," *Consequence of Sound*, November 14, 2013, https://consequenceofsound.net/2013/11/album-review-mount-eerie-pre-human -ideas/.

30. In Daniel Mescher, "Phil Elverum of Mount Eerie: 'Songs Are Evolving, Living Things,'" *Colorado Public Radio* (Open Air), October 8, 2014, https://www.cpr.org/ 2014/10/08/phil-elverum-of-mount-eerie-songs-are-evolving-living-things/.

31. Ibid.

32. Mount Eerie, *Wind's Poem*, Elverum & Sun, 2009, CD.

33. Mount Eerie, *A Crow Looked at Me*, Elverum & Sun, 2017, CD.

34. Mount Eerie, *Now Only*, Elverum & Sun, 2018, CD.

35. Sam Kidel, *Disruptive Muzak*, The Death of Rave, 2016, YouTube audio.

36. Paul Schwartzman, "I Gave a Famous Rock Star a Windshield Tour of D.C.— and Didn't Know Who She Was," *Washington Post*, March 18, 2016, https://www .washingtonpost.com/lifestyle/i-gave-a-famous-rock-star-a-windshield-tour -of-dc--and-didnt-know-who-she-was/2016/03/18/8124ab28-e488-11e5-bc08 -3e03a5b41910_story.html.

37. PJ Harvey, *The Hope Six Demolition Project*, Island/Vagrant, 2016, CD.

38. "PJ Harvey: Recording in Progress," *Somerset House*, 2015, https://www .somersethouse.org.uk/whats-on/pj-harvey-recording-in-progress.

39. Ibid.

40. John D. Barrow, *The Artful Universe* (New York: Oxford University Press, 1995), 190.

41. "Washington DC to PJ Harvey: 'She's to Music What Piers Morgan Is to Cable News,'" *The Guardian*, March 17, 2016, https://www.theguardian.com/music/

2016/mar/17/pj-harvey-washington-community-of-hope-six-demolition-project
-ward-7.

42. "PJ Harvey The Community of Hope Andrew Marr Show 2016," 2016, YouTube
video.

43. Kanye West, *The Life of Pablo*, GOOD/Def Jam, 2016, Streaming audio.

44. Joe Coscarelli, "400 Million Streams Later, Kanye West's 'Pablo' Gets a Wider
Release," *New York Times*, March 31, 2016, https://www.nytimes.com/2016/04/02/
arts/music/kanye-west-life-pablo-tidal-streams.html.

45. Toyomu, *Imagining the Life of Pablo*, self-released, 2016, YouTube audio.

46. Reuben Tasker, "Talking with Toyomu, the Producer Who Imagined 'The Life of
Pablo' for a Tidal-Less Japan," *Genius*, April 8, 2016, https://genius.com/a/talking
-with-toyomu-the-producer-who-imagined-the-life-of-pablo-for-a-tidal-less
-japan.

47. Any apparent grammatical errors in this quotation are likely the result of a
language/translation difference between the interviewer and interviewee.

48. The Fiery Furnaces, *Rehearsing My Choir*, Rough Trade, 2005, CD.

49. Matthew Friedberger, "From the Artist," *Amazon* (*Rehearsing My Choir* product
page), October 25, 2005, https://www.amazon.com/Rehearsing-My-Choir-Fiery
-Furnaces/dp/B000BDJ02U.

50. The Fiery Furnaces, *Rehearsing My Choir*.

51. Ibid.

52. Bergen Evans, *The Natural History of Nonsense* (New York: Knopf, 1946), 10.

53. "Coditany of Timeness," 2017, *Bandcamp*, https://dadabots.bandcamp.com/album/
coditany-of-timeness.

# Bibliography


Abbott, Stacey. "The Battlefield for the Soul: Special Effects and the Possessed Body." In *Special Effects: New Histories/Theories/Contexts*, edited by Dan North, Bob Rehak, and Michael S. Duffy, 141–53. London: BFI Publishing, 2015.

Adorno, Theodor, and Max Horkheimer. *Dialectic of Enlightenment*. Stanford, CA: Stanford University Press, 2002.

Aiken, Lewis R. *Human Development in Adulthood*. New York: Plenum Press, 1998.

Alfred, Ruth Ann. "The Effect of Censorship on American Film Adaptations of Shakespearean plays." Master's thesis, Texas A&M University, 2008.

Altman, Rick. *A Theory of Narrative*. New York: Columbia University Press, 2008.

*Amor de Perdição* (censored screenplay). António Lopes Ribeiro, 1942. Portuguese Film Library archives.

Andrew, Dudley. "Adaptation." In *Film Adaptation*, edited by James Naremore, 28–37. New Brunswick, NJ: Rutgers University Press, 2000.

———. *Concepts in Film Theory*. Oxford: Oxford University Press, 1984.

António, Lauro. *Cinema e censura em Portugal: 1926–1974*. Lisbon: Arcádia, 1978.

Antonio, Sheril D. *Contemporary African American Cinema*. New York: Peter Lang, 2002.

Appadurai, Arjun. *Modernity at Large: Cultural Dimensions of Globalization*. Minneapolis: University of Minnesota Press, 1996.

Arnett, Jeffrey Jensen. *Emerging Adulthood: The Winding Road from the Late Teens through the Twenties*. Oxford: Oxford University Press, 2004.

*As Pupilas do Senhor Reitor* (censored screenplay and file). Perdigão Queiroga, 1959. Portuguese Film Library archives.

Athanasourelis, John Paul. "Film Adaptation and the Censors: 1940's Hollywood and Raymond Chandler." *Studies in the Novel* 35 (2003): 325–38.

Baldi, Pierre. *The Shattered Self: The End of Natural Evolution*. Cambridge, MA: MIT Press, 2002.

Barrow, John D. *The Artful Universe*. New York: Oxford University Press, 1995.

Barthes, Roland. *The Rustle of Language*. Trans. Richard Howard. Berkeley: University of California Press, [1967] 1989.

———. *S/Z*. Paris: Seuil, 1970.

Basinski, William. *The Disintegration Loops*. 2002–3. 2062, CD.

Battaglia, Andy. "From Destruction, a New Sound of Life." *Wall Street Journal*, September 10, 2012. https://www.wsj.com/articles/SB1000087239639044 45547045776436722373386032.

Batty, Craig. *Movies That Move Us*. Basingstoke, UK: Palgrave Macmillan, 2011.

Bauman, Zygmunt. *Globalization: The Human Consequences*. Cambridge: Polity, 1998.

Beauchamp, Scott. "How Vaporwave Was Created Then Destroyed by the Internet." *Esquire*, August 18, 2016. https://www.esquire.com/entertainment/music/a47793/what-happened-to-vaporwave/.

Berman, Eliza. "Everything to Know about the *Shape of Water* Plagiarism Controversy." *Time*, March 3, 2018. http://time.com/5170613/shape-of-water-plagiarism-controversy/.

Biltereyst, Daniel, and Roel Vande Winkel. *Silencing Cinema: Film Censorship around the World*. New York: Palgrave Macmillan, 2013.

Birnbaum, Debra. "Freddie Highmore on 'Bates Motel' Series Finale: 'It's the Fitting End to a Love Story.'" *Variety*, April 24, 2017. https://variety.com/2017/tv/news/freddie-highmore-bates-motel-finale-season-5-episode-10-1202392130/.

Birzer, Bradley J. *J.R.R. Tolkien's Sanctifying Myth: Understanding Middle-earth*. Wilmington, DE: ISI Books, 2003.

Blanchet, Alexis. "A Statistical Analysis of the Adaptation of Films into Video Games." *Inaglobal.fr*, December 7, 2011. http://www.inaglobal.fr/en/video-games/article/statistical-analysis-adaptation-films-video-games. Accessed March 20, 2017.

Blatterer, Harry. *Coming of Age in Times of Uncertainty*. New York: Berghahn Books, 2007.

Bluestone, George. *Novels into Film*. Baltimore: Johns Hopkins University Press, [1957] 2003.

Bobo, Jacqueline. "Reading Through the Text: The Black Woman as Audience." In *Black American Cinema*, edited by Manthia Diawara, 272–87. New York: Routledge, 1993.

Bondanella, Peter. *The Films of Federico Fellini*. Cambridge: Cambridge University Press, 2002.

Boozer, Jack. *Authorship in Film Adaptation*. Austin: University of Texas Press, 2008.

———. "Introduction: The Screenplay and Authorship in Adaptation." In *Authorship in Film Adaptation*, edited by Jack Boozer, 1–30. Austin: University of Texas Press, 2008.

Braudy, Leo. "Afterword: Rethinking Remakes." In *Play It Again, Sam: Retakes on Remakes*, edited by Andrew Horton and Stuart Y. McDougal, 327–34. Berkeley: University of California Press, 1998.

Breuklander, Joel. "Straight, White Guys, At Least." *The Atlantic*, July 18, 2013. https://www.theatlantic.com/entertainment/archive/2013/07/literature-is -dead-according-to-straight-white-guys-at-least/277906/.

Brooker, Will. *Hunting the Dark Knight: Twenty First Century Batman*. London: I. B. Tauris, 2012.

Brown, Dan. *The Da Vinci Code*. London: Corgi Books, 2003.

Brown, Simon. "Alternate Versions of the Same Reality: Adapting *Under the Dome* as an SF Television Series." *Science Fiction Film and Television* 10, no. 2 (2017): 267–83.

———. "Memento Mori: The Slow Death of *The X-Files*." *Science Fiction Film and Television* 6, no. 1 (2013): 7–22.

———. *Screening Stephen King: Adaptation and the Horror Genre in Film and Television*. Austin: University of Texas Press, 2018.

Bruner, Jerome. "Life as Narrative." *Social Research: An International Quarterly* 71, no. 3 (2004): 691–710.

Brütsch, Matthias. "The Three-Act Structure: Myth or Magical Formula?" *Journal of Screenwriting* 6, no. 3 (2015): 301–26.

Bubeníček, Petr. *Subversive Adaptations: Czech Literature on Screen behind the Iron Curtain*. London: Palgrave Macmillan, 2017.

Burke, Liam. *The Comic Book Film Adaptation: Exploring Modern Hollywood's Leading Genre*. Jackson: University of Mississippi Press, 2016.

*Camoëns*. By Leitão de Barros (reference FIFR1-B1). Cannes International Film Festival Fund, "*service régie*," program and press release, 1946, French Film Library Archives.

Campbell, Joseph. *The Hero with a Thousand Faces*. 3rd ed. Novato: New World Library, [1949] 2008.

———. *The Power of Myth*. New York: Doubleday, 1988.

Canby, Vincent. "The Screen: Westworld." *New York Times*, November 22, 1973. https://www.nytimes.com/1973/11/22/archives/the-screen-westworld-robots -and-fantasies-in-film-by-crichton-the.html.

Cardullo, Bert. *European Directors and Their Films: Essays on Cinema*. Lanham, MD: Scarecrow Press, 2012.

Cardullo, Bert, and Alain Piette. *Bazin at Work: Major Essays & Reviews from the Forties and Fifties*. New York: Routledge, 1997.

Cardwell, Sarah. "Is Quality Television Any Good?" In *Quality TV: Contemporary American Television and Beyond*, edited by Janet McCabe and Kim Akass, 19–34. London: I. B. Tauris, 2007.

———. *Adaptation Revisited: Television and the Classic Novel*. Manchester: Manchester University Press, 2002.

Cartmell, Deborah, and Imelda Whelehan, eds. *Adaptations: From Text to Screen, Screen to Text*. New York: Routledge, 1999.

Cattrysse, Patrick. *Descriptive Adaptation Studies: Epistemological and Methodological Issues*. Antwerp: Garant, 2014.

Caygill, Howard. *Walter Benjamin: The Colour of Experience*. New York: Routledge, 1998.

Chan, Kenneth. "The Chinese Cinematic Remake as Transnational Appeal: Zhang Yimou's *A Woman, a Gun and a Noodle Shop*." In *Transnational Film Remakes*, edited by Iain Robert Smith and Constantine Verevis, 87–102. Edinburgh: Edinburgh University Press, 2017.

Chandra, Anupama. "Cool Copy Cats." *India Today*, August 1993, 76–77.

Chappel, Jacquelyn. "Lost in Translation." *New York Times Book Review*, September 11, 2016. https://www.nytimes.com/2016/09/11/books/review/letters-to -theeditor.html.

Clüver, Claus. "Ekphrasis and Adaptation." In Leitch, *Oxford Handbook of Adaptation Studies*, 459–76.

"Coditany of Timeness." 2017. *Bandcamp*. https://dadabots.bandcamp.com/album/ coditany-of-timeness.

Collins, Steve. "Amen to That: Sampling and Adapting the Past." *M/C Journal* 10, no. 2 (2007). http://journal.media-culture.org.au/0705/09-collins.php.

Constandinides, Costas. *From Film Adaptation to Post-Celluloid Adaptation*. New York: Continuum, 2010.

Corrigan, Timothy. "Defining Adaptation." In Leitch, *Oxford Handbook of Adaptation Studies*, 23–35.

Coscarelli, Joe. "400 Million Streams Later, Kanye West's 'Pablo' Gets a Wider Release." *New York Times*, March 31, 2016. https://www.nytimes.com/2016/04/ 02/arts/music/kanye-west-life-pablo-tidal-streams.html.

Côté, James A. *Arrested Adulthood: The Changing Nature of Maturity and Identity*. New York: New York University Press, 2000.

Cowan, Jane K. *Dance and the Body Politic in Northern Greece*. Princeton: Princeton University Press, 1990.

Cutchins, Dennis. "Bakhtin, Intertextuality, and Adaptation." In Leitch, *Oxford Handbook of Adaptation Studies*, 71–87.

Davis, Darrell William. "Reigniting Japanese Tradition with 'Hana-Bi.'" *Cinema Journal* 40, no. 4 (2001): 55–80.

Della Coletta, Cristina. *When Stories Travel: Cross-Cultural Encounters between Fiction and Film*. Baltimore: Johns Hopkins University Press, 2012.

Delveroudi, Eliza-Anna. *Oi Neoi stis Komodies tou Ellinikou Kinimatografou 1948–1974 / Youth in Greek Cinema Comedies, 1948–1974*. Athens: IAEN/INE, 2004.

Desta, Yohana. "Guillermo del Toro Vigorously Defends *Shape of Water* against Plagiarism Claim." *Vanity Fair*, February 22, 2018. https://www.vanityfair .com/hollywood/2018/02/guillermo-del-toro-defends-shape-of-water-lawsuit.

Diaz, Eric. "*The Exorcist* Being Remade as a Pilot for Fox." *Nerdist*, January 23, 2016. https://nerdist.com/the-exorcist-being-remade-as-pilot-for-fox/. Accessed April 15, 2018.

Donoghue, Emma. *Room*. London: Picador, 2010.

Dove, Steve. "Asghar Farhadi Oscar 2017 Winner Speech Delivered by Anousheh Ansari." *Oscar.go.com*, February 27, 2017. http://oscar.go.com/news/winners/ asghar-farhadi-oscar-2017-winner-speech-delivered-by-anousheh-ansari.

Dowd, Vincent. "Oscars: Selma Writer Tells His Side of Row with Director." *BBC*, February 20, 2015. https://www.bbc.com/news/entertainment-arts-31539526.

Dutton, Denis. *The Art Instinct*. Oxford: Oxford University Press, 2009.

Eco, Umberto. *Reflections on* The Name of the Rose. Translated by William Weaver. London: Minerva, 1994.

———. *The Role of the Reader: Explorations in the Semiotics of Texts*. Bloomington: Indiana University Press, 1979.

Edgerton, Gary R., and Brian G. Rose, eds. *Thinking Outside the Box: A Contemporary Television Genre Reader*. Lexington: University Press of Kentucky, 2005.

Elleström, Lars. "Adaptation and Intertextuality." In Leitch, *Oxford Handbook of Adaptation Studies*, 509–26.

Elliot, Kamilla. "How Do We Talk about Adaptation Studies Today?" *Literature/ Film Quarterly* 45, no. 2 (2017). https://lfq.salisbury.edu/_issues/first/how_do _we_talk_about_adaptation_studies_today.html.

———. "Literary Film Adaptation and the Form/Content Dilemma." In *Narratives Across Media*, edited by Marie-Laure Ryan, 220–43. Nebraska: University of Nebraska Press, 2004.

———. *Rethinking the Novel/Film Debate*. Cambridge: Cambridge University Press, 2003.

Ellis, John. "The Literary Adaptation." *Screen* 23, no.1 (1982): 3–5.

Erikson, Erik H. *Identity and the Life Cycle*. New York: Norton, [1959] 1980.

Evans, Bergen. *The Natural History of Nonsense*. New York: Knopf, 1946.

Eyerman, Ron. *Cultural Trauma: Slavery and the Formation of African American Identity*. Cambridge: Cambridge University Press, 2001.

Ferro, Antonio. *Salazar, le Portugal et son chef*. Paris: Editions Bernard Grasset, 1934.

———. *Teatro e cinema, 1936–1949*. Lisbon: Editions SNI, 1950.

Ferro, Marc. *Cinéma et Histoire*. Paris: Gallimard, 2005.

Feuer, Jane. "Is Dirty Dancing a Musical, and Why Should It Matter?" In *The Time of Our Lives: "Dirty Dancing" and Popular Culture*, edited by Yannis Tzioumakis and Sian Lincoln, 59–72. Detroit, MI: Wayne State University Press, 2013.

The Fiery Furnaces. *Rehearsing My Choir*. Rough Trade, 2005. CD.

Fisher, Mark. *Ghosts of My Life: Writings on Depression, Hauntology and Lost Futures*. Hants, UK: Zero Books, 2014.

Flynn, Gillian. *Gone Girl*. London: Phoenix, 2012.

Foucault, Michel. *Dits et Écrits (1954–1988)*. Vol. 1. Paris: Gallimard, 2001.

Francis, Terri. "'She Will Never Look': Film Spectatorship, Black Feminism and Scary Subjectivities." In *Reclaiming the Archive*, edited by Vicki Callahan, 98–125. Detroit, MI: Wayne State University Press, 2011.

Friedberger, Matthew. "From the Artist." *Amazon* (*Rehearsing My Choir* product page). October 25, 2005. https://www.amazon.com/Rehearsing-My-Choir -Fiery-Furnaces/dp/B000BDJ02U.

Fuller, Graham. "Of Human Bondage." *Film Comment* 49, no. 5: 20. MasterFILE Premier, EBSCOhost, 2013.

Gallix, Andrew. "Hauntology: A Not-So-New Critical Manifestation." *The Guardian*, June 17, 2011. https://www.theguardian.com/books/booksblog/2011/jun/ 17/hauntology-critical.

Genette, Gérard. *Narrative Discourse: An Essay in Method*. Translated by Jane E. Lewin. New York: Cornell University Press, 1980.

———. *Palimpsestes: La littérature au second degré*. Paris: Seuil, 1982.

Geraghty, Christine. *Now a Major Motion Picture: Film Adaptations of Literature and Drama*. Lanham, MD: Rowman & Littlefield, 2007.

Gibson, Caitlin. "How Accurate Is 'Selma'?" *Washington Post*, February 17, 2015. https://www.washingtonpost.com/news/arts-and-entertainment/wp/2015/02/ 17/how-accurate-is-selma/?utm_term=.bbf14bbdc702. Accessed March 20, 2017.

Gizelis, Grigoris, Roxani Kaftantzoglou, Afroditi Teperoglou, and Vasilis Filias. *Paradosi kai Neoterikotita stis Politistikes Drastiriotites tis Ellinikis Oikogeneias / Tradition and Postmodernims in the Cultural Activities of the Greek Family*. Athens: EKKE, 1984.

Gjelsvik, Anne. "What Novels Can Tell That Movies Can't Show." In *Adaptation Studies: New Challenges, New Directions*, edited by Jorgen Bruhn, Anne Gjelsvik, and Eirik Frisvold Hanssen, 245–63. Bloomsbury: London, 2013.

Grant, Barry Keith, and Scott Henderson, eds. *Comics and Pop Culture: Adaptation from Panel to Frame*. Austin: University of Texas Press, 2019.

Gravdal, Kathryn. *Ravishing Maidens: Writing Rape in Medieval French Literature and Law*. Philadelphia: University of Pennsylvania Press, 1991.

Gray, Brandon. "'Gulliver,' 'Persia,' 'Narnia' Rank among the Big Botches of 2010." *Boxofficemojo*, January 20, 2011. http://www.boxofficemojo.com/news/?id=3051&p=.htm.

Gray, Tim. "Love and Danger on the 'Water' Front." *Variety*, January 10, 2018. https://variety.com/2018/film/awards/shape-of-water-inspiration-from-monster-movie-1202659976/.

Greimas, Algirdas, J. *Sémantique structurale*. Paris: Larousse, 1966.

Greimas, Algirdas, J., and Joseph Courtés. *Sémiotique: Dictionnaire raisonné de la théorie du langage*. Paris: Hachette, 1986.

Griggs, Yvonne. *Adaptable TV*. New York: Palgrave, 2018.

———. *The Bloomsbury Introduction to Adaptation Studies: Adapting the Canon in Film, TV, Novels and Popular Culture*. New York: Bloomsbury, 2016.

Grossman, Julie. *Literature, Film, and Their Hideous Progeny: Adaptation and Elas-TEXTity*. New York: Palgrave, 2015.

Grossman, Julie, and R. Barton Palmer, eds. *Adaptation in Visual Culture: Images, Texts, and Their Multiple Worlds*. New York: Palgrave, 2017.

Hand, Richard J., and Andrew Purssell. *Adapting Graham Greene*. New York: Red Globe Press, 2015.

Harvey, PJ. *The Hope Six Demolition Project*. Island/Vagrant, 2016. CD.

Hayward, Susan. *Key Concepts in Cinema Studies*. London: Routledge, 1996.

Hertzberg, Hendrik. "'Lincoln' v. Lincoln." *New Yorker*, December 17, 2012. https://www.newyorker.com/news/hendrik-hertzberg/lincoln-v-lincoln.

Hilmes, Michele. "The Whole World's Unlikely Heroine: Ugly Betty as Transnational Phenomenon." In *Reading Ugly Betty: TV's Betty Goes Global*, edited by Janet McCabe and Kim Akass, 26–44. London: I. B. Tauris, 2013.

Hogan, Patrick Colm. *Affective Narratology: The Emotional Structure of Stories*. Lincoln: University of Nebraska Press, 2011.

Hollows, Stephen. "How Original Are Hollywood Movies?" *Stephenhollows. com*, June 8, 2015. https://stephenfollows.com/how-original-are-hollywood -movies/.

Holzer, Harold. "What's True and False in 'Lincoln' Movie." *Daily Beast*, November 22, 2012. https://www.thedailybeast.com/author/harold-holzer.

hooks, bell. "The Oppositional Gaze: Black Female Spectators." In *Black American Cinema*, edited by Manthia Diawara, 288–302. New York: Routledge, 1993.

Horeck, Tanya. *Public Rape: Representing Violation in Fiction and Film*. London: Routledge, 2004.

Horta e Costa, António. *Subsídios para a história do cinema português: 1896–1949*. Lisbonne: Empresa Literaria Universal, 1949.

Hutcheon, Linda. *A Theory of Adaptation*. 2nd ed. Oxon: Routledge, 2013.

James, E. L. *Fifty Shades of Grey*. New York: Vintage Books, 2012.

Jeffers, Jennifer L. *Britain Colonized: Hollywood's Appropriation of British Literature*. New York: Palgrave Macmillan, 2006.

Jeffries, Dru. *Comic Book Film Style: Cinema at 24 Panels per Second*. Austin: University of Texas Press, 2017.

Jellenik, Glenn. "On the Origins of Adaptation, as Such: The Birth of a Simple Abstraction." In Leitch, *Oxford Handbook of Adaptation Studies*, 36–52.

Jenkins, Henry. *Convergence Culture: Where Old and New Media Collide*. New York: New York University Press, 2006.

———. *Textual Poachers: Television Fans and Participatory Culture*. New York: Routledge, 1992.

Jess-Cooke, Carolyn, and Constantine Verevis. "Introduction." In *Second Takes: Critical Approaches to the Film Sequel*, edited by Carolyn Jess-Cooke and Constantine Verevis, 1–10. Albany: SUNY Press, 2010.

———. *Second Takes: Critical Approaches to the Film Sequel*. Albany: SUNY Press, 2010.

Jones, Jacquie. "The Construction of Black Sexuality: Towards Normalizing the Black Cinematic Experience." In *Black American Cinema*, edited by Manthia Diawara, 247–56. New York: Routledge, 1993.

Jowett, Lorna, and Stacey Abbott. *TV Horror: Investigating the Dark Side of the Small Screen*. London: I. B. Tauris, 2013.

———. "TV Horror: *Santa Clarita Diet*." In *Horror Reader*, edited by Simon Bacon, 45–52. Bern: Peter Lang, 2019.

Kaklamanidou, Betty. *The "Disguised" Political Film in Contemporary Hollywood: A Genre's Construction*. New York: Bloomsbury, 2016.

Kassaveti, Orsalia-Eleni. *I Elliniki Videotainia (1985–1990): Eidologikes, Koinonikes & Politismikes Diastaseis / The Greek Video-Film (1985–1990): Generic, Social and Cultural Aspects*. Athens: Asini, 2014.

———. "To Elliniko Melodrama. I Ekseliksi enos Dimofilous Kinimatografikou Eidous." In *Apo ton Proimo ston Syghrono Elliniko Kinimatografo / From Early to Contemporary Greek Cinema*, edited by Maria Paradeisi and Afroditi Nikolaidou, 45–75. Athens: Gutenberg, 2017.

Keegan, Rebecca. "How Guillermo del Toro Crafted the Perfect Monster-Romance for the Trump Era." *Vanity Fair*, September 4, 2017. https://www.vanityfair.com/hollywood/2017/09/guillermo-del-toro-shape-of-water.

Kelleter, Frank, and Kathleen Loock. "Hollywood Remaking as Second-Order Serialization." In *Media of Serial Narrative*, edited by Frank Kelleter, 125–47. Columbus: Ohio State University Press, 2017.

Kidel, Sam. *Disruptive Muzak*. The Death of Rave, 2016. YouTube audio.

Kiefer, Halle. "*Game of Thrones, The Walking Dead* and *Westworld* Are the Most Torrented Shows of 2016." *Vulture*, December 27, 2016. http://www.vulture.com/2016/12/game-of-thrones-westworld-walking-dead-most-torrented-tv-2016.html.

King, Stephen D. *Grave New World: The End of Globalization, the Return of History*. London: Yale University Press, 2017.

Kirn, Walter. *Up in the Air*. New York: Doubleday, 2001.

Kit, Borys. "How Guillermo del Toro's 'Black Lagoon' Fantasy Inspired 'Shape of Water.'" *Hollywood Reporter*, November 3, 2017. https://www.hollywoodreporter.com/news/how-guillermo-del-toros-black-lagoon-fantasy-inspired-shape-water-1053206.

Konstam, Varda. *Emerging and Young Adulthood: Multiple Perspectives, Diverse Narratives*. Heidelberg: Springer, 2015.

Kornetis, Kostis. *Children of the Dictatorship: Student Resistance, Cultural Politics and the "Long 1960s" in Greece*. New York: Berghahn, 2013.

Koski, Genevieve, Keith Phipps, and Tasha Robinson. "The Problem with 'Based on a True Story.'" *The Dissolve*, September 16, 2013. https://thedissolve.com/features/the-conversation/156-the-problem-with-based-on-a-true-story/.

Kouanis, Panos. *I Kinimatografiki Agora stin Ellada (1944–1999) / Film Market in Greece (1944–1999)*. Athens: Finatec, 2001.

Krallice. *Years Past Matter*. Self-released, 2012. CD.

Krämer, Lucia. "Adaptation in Bollywood." In Leitch, *Oxford Handbook of Adaptation Studies*, 251–66.

Kupfer, Joseph H. *Visions of Virtue in Popular Film*. Boulder, CO: Westview Press, 1999.

Lang, Brent. "Global Box Office Hits Record $40.6 Billion in 2017; U.S. Attendance Lowest in 23 Years." *Variety*, April 4, 2018. https://variety.com/2018/digital/news/global-box-office-hits-record-40-6-billion-in-2017-u-s-attendance-lowest-in-23-years-1202742991/.

Larson, Jeremy D. "Mount Eerie: *Pre-Human Ideas*." *Pitchfork*, November 11, 2013. https://pitchfork.com/reviews/albums/18719-mount-eerie-pre-human-ideas/.

Lawler, Kelly. "Has 'The Walking Dead' Finally Gone Too Far for Fans?" *USA Today*, October 24, 2016. https://eu.usatoday.com/story/life/entertainthis/2016/10/24/walking-dead-season-7-premiere-fan-reaction/92669814/.

Lehane, Dennis. *Shutter Island*. London: Bantam Press, 2003.

Leitch, Thomas. "Adaptation and Intertextuality, or, What Isn't an Adaptation, and What Does It Matter?" In *A Companion to Literature, Film and Adaptation*, edited by Deborah Cartmell, 87–102. Hoboken, NJ: Wiley-Blackwell, 2012.

———. "Against Conclusions: Petit Theories and Adaptation Studies." In Leitch, *Oxford Handbook of Adaptation Studies*, 698–711.

———. *Film Adaptation and Its Discontents*. Baltimore: Johns Hopkins University Press, 2007.

———. "Introduction." In Leitch, *Oxford Handbook of Adaptation Studies*, 1–22.

———. "Mind the Gaps." In *Adaptation in Visual Culture: Images, Texts, and Their Worlds*, edited by Julie Grossman and Barton Palmer, 53–71. London: Palgrave, 2017.

———, ed. *The Oxford Handbook of Adaptation Studies*. Oxford: Oxford University Press, 2017.

———. "The Texts behind *The Killers*." In *Twentieth-Century American Fiction on Screen*, edited by R. Barton Palmer. 27–46. Cambridge: Cambridge University Press, 2007.

———. "Twelve Fallacies in Contemporary Adaptation Theory." *Criticism* 45, no. 2 (2003): 149–71.

———. "Twice Told Tales: The Rhetoric of the Remake." *Literature/Film Quarterly* 18, no. 3 (1990): 138–49.

Lev, Peter. "How to Write Adaptation History." In Leitch, *Oxford Handbook of Adaptation Studies*, 661–78.

Lockett, Dee. "How Accurate Is *Selma*?" *Slate*, December 14, 2014. http://www.slate.com/blogs/browbeat/2014/12/24/selma_fact_vs_fiction_how_true_ava_duvernay_s_new_movie_is_to_the_1965_marches.html?via=gdpr-consent.

Logan, Elliott. *Breaking Bad and Dignity: Unity and Fragmentation in the Serial Television Drama*. New York: Palgrave Macmillan, 2016.

Loock, Kathleen. "The Past Is Never Really Past: Serial Storytelling from *Psycho* to *Bates Motel*." *Literatur in Wissenschaft und Unterricht* 18, nos. 1–2 (2014): 81–96.

Loock, Kathleen, and Constantine Verevis, eds. *Film Remakes, Adaptations and Fan Productions*. New York: Palgrave, 2012.

Luhr, William. "Adapting *Farewell, My Lovely*." In *A Companion to Literature and Film*, edited by Robert Stam and Alessandra Raengo, 278–97. Malden, MA: Blackwell, 2004.

Macintosh Plus. *Floral Shoppe*. Beer on the Rug, 2011. YouTube audio.

Maltby, Richard. "'To Prevent the Prevalent Type of Book': Censorship and Adaptation in Hollywood, 1924–1934." *American Quarterly* 44, no. 4 (1992): 554–83.

Martin, Cathlena. "Charlotte's Website: Media Transformation and the Intertextual Web of Children's Culture." In *Adaptation in Contemporary Culture: Textual Infidelities*, edited by Rachel Carroll, 85–95. London: Continuum, 2009.

Martin, George R. R. *A Song of Ice and Fire*. Vols. 1–7. London: Harper Voyager, 1996.

Martinez, Kiko. "James D. Solomon: *The Conspirator*." *CineSnob*, April 29, 2011. http://www.cinesnob.net/james-d-solomon-the-conspirator/.

McAdams, Dan. *The Redemptive Self: Stories Americans Live By*. Oxford: Oxford University Press, 2006.

McCabe, Janet, and Kim Akass, eds. *Quality TV: Contemporary American Television and Beyond*. London: I. B. Tauris, 2007.

McCallum, Robyn. *Screen Adaptations and the Politics of Childhood*. New York: Palgrave, 2018.

McCarthy, Cormac. *The Road*. New York: Random House, 2007.

McEwan, Ian. *Atonement*. London: Jonathan Cape, 2001.

McFarland, Douglas, and Wesley King, eds. *John Huston as Adaptor*. Albany: SUNY Press, 2018.

McFarlane, Brian. *Novel to Film*. Oxford: Clarendon Press, 1996.

McLean, Jesse. *The Art and Making of Hannibal*. London: Titan Books, 2015.

Mescher, Daniel. "Phil Elverum of Mount Eerie: 'Songs Are Evolving, Living Things.'" *Colorado Public Radio* (Open Air), October 8, 2014. https://www.cpr.org/2014/10/08/phil-elverum-of-mount-eerie-songs-are-evolving-living-things/.

Meyer, Michael J., ed. *Literature and Musical Adaptation*. Amsterdam: Rodopi, 2002.

Miller, Julie. "How Greta Gerwig's *Lady Bird* Came to 'Look Like a Memory.'" *Vanity Fair*, November 3, 2017. https://www.vanityfair.com/hollywood/2017/11/greta-gerwig-lady-bird-design.

Miller, Toby. *Television Studies: The Basics*. New York: Routledge, 2010.

Mitaine, Benoit, et al. *Comics and Adaptation*. Jackson: University of Mississippi Press, 2016.

Mittell, Jason. *Complex TV: The Poetics of Contemporary Television Storytelling*. New York: New York University Press, 2015.

———. *Genre and Television: From Cop Shows to Cartoons in American Culture*. New York: Routledge, 2004.

Molla, Rani. "Netflix Now Has Nearly 118 Million Streaming Subscribers Globally." *Recode*, January 22, 2018. https://www.recode.net/2018/1/22/16920150/netflix-q4-2017-earnings-subscribers.

Motion Picture Association of America (MPPA). "Theme Report, 2017." https://pmcdeadline2.files.wordpress.com/2018/04/mpaa-theme-report-2017.pdf.

Mount Eerie. *A Crow Looked at Me*. P. W. Elverum & Sun, 2017. CD.

———. *Now Only*. P. W. Elverum & Sun, 2018. CD.

———. *Pre-Human Ideas*. P. W. Elverum & Sun, 2013. YouTube audio.

———. *Wind's Poem*. Elverum & Sun, 2009. CD.

Mouzelis, Nicos P. *Modern Greece: Facets of Underdevelopment*. London: MacMillan Press, 1978.

Moyes, Jojo. 2012. *Me Before You*. London: Penguin.

Murray, Simone. *The Adaptation Industry: The Cultural Economy of Contemporary Literary Adaptation*. London: Routledge, 2012.

———. "The Business of Adaptation: Reading the Market." In *A Companion to Literature, Film, and Adaptation*, edited by Deborah Cartmell, 122–40. Malden, MA: Wiley-Blackwell, 2012.

Naremore, James., ed. *Film Adaptation*. New Brunswick, NJ: Rutgers University Press, 2000.

———. "Introduction: Film and the Reign of Adaptation." In *Film Adaptation*, edited by James Naremore, 1–18. New Brunswick, NJ: Rutgers University Press, 2000.

Nayar, Sheila J. "The Values of Fantasy: Indian Popular Cinema through Western Scripts." *Journal of Popular Culture* 31, no. 1 (1997): 73–90.

Ndounou, Monica W. *Shaping the Future of African American Films: Color-Coded Economics and the Story behind the Numbers*. New Brunswick, NJ: Rutgers University Press, 2014.

Neale, Steve. *Genre & Hollywood*. London: Routledge, 2000.

Nededog, Jethro. "'Walking Dead' Creator Explains His Original Vision for the Series." *Business Insider UK*, October 9, 2015. http://uk.businessinsider.com/walking-dead-creator-explains-his-original-vision-for-the-series-2015-10.

Newcomb, Henry. "Reflections on *TV: The Most Popular Art*." In *Thinking Outside the Box: A Contemporary Television Genre Reader*, edited by Gary R. Edgerton and Brian G. Rose, 17–36. Lexington: University Press of Kentucky, 2005.

Nikolakakis, Michalis. "Tourism, Body and Seaside Recreational Practices in Post-war Greek Society Until 1974." In *Consumption & Gender in Southern Europe since the Long 1960s*, edited by Kostis Kornetis, Eleni Kotsovili, and Nikos Papadogiannis, 103–17. London: Bloomsbury, 2016.

Northup, Solomon. *Twelve Years a Slave*. Baton Rouge: Louisiana State University Press, 1968.

Office of National Information Fund. *Direção Geral dos Espetáculos* Fund. National Archives of Torre do Tombo, Lisbon, Portugal (SNI IGAC, ANTT).

Osterhammel, Jürgen, and Niels P. Petersson. *Globalization: A Short History*. New Brunswick, NJ: Princeton University Press, 2009.

O'Sullivan, Sean. "*The Sopranos*: Episodic Storytelling." In *How to Watch Television*, edited by Ethan Thompson and Jason Mittell, 65–73. New York: New York University Press, 2013.

Panofsky, Erwin. "The Concept of Artistic Volition." Trans. Kenneth J. Northcott and Joel Snyder. *Critical Inquiry* 8, no. 1 (1981): 17–33.

Papadimitriou, Lydia. *To Elliniko Kinimatografiko Musical / Greek Film Musical*. Athens: Papazisi, 2009.

Papazian Gretchen, and Joseph Michael Sommers, eds. *Game On, Hollywood! Essays on the Intersection of Video Games and Cinema*. Jefferson, NC: McFarland, 2013.

Paradeisi, Maria. "Erga kai Imerai tou Paliou kai tou Neou Ellinikou Kinimato-grafou" / "Works of and Old and New Greek Cinema." *O Politis* 122 (1993): 50–56.

Pearson, Roberta. "*Lost* in Transition: From Post-Network to Post-Television." In *Quality TV: Contemporary American Television and Beyond*, edited by Janet McCabe and Kim Akass, 239–56. London: I. B. Tauris, 2007.

Peary, Gerald, and Roger Shatzkin, eds. *The Classic American Novel and the Movies*. New York: Ungar, 1977.

———. *The Modern American Novel and the Movies*. New York: Ungar, 1978.

Perry, Dennis R. "The Recombinant Mystery of Frankenstein: Experiments in Film Adaptation." In Leitch, *Oxford Handbook of Adaptation Studies*, 137–53.

Peterson, Jeff. "Why Peter Jackson's *Lord of the Rings* Succeeded as an Adaptation." *Deseret News*, December 7, 2012. https://www.deseretnews.com/article/865568286/Why-Peter-Jacksons-Lord-of-the-Rings-succeeded-as-an-adaptation.html.

Phelan, James. *Reading People, Reading Plots: Character, Progression, and the Interpretation of Narrative*. Chicago: University of Chicago Press, 1989.

Phelan, Peggy. *Unmarked: The Politics of Performance*. London: Routledge, 1993.

Piçarra, Maria do Carmo. *Azuis ultramarinos, Propaganda Colonial e Censura no Cinema do Estado Novo*. Lisbon: Edições 70, 2015.

"PJ Harvey The Community of Hope Andrew Marr Show 2016." 2016. YouTube video.

"PJ Harvey: Recording in Progress." *Somerset House*. 2015. https://www.somersethouse.org.uk/whats-on/pj-harvey-recording-in-progress.

Pym, Anthony D. *The Moving Text: Localization, Translation, and Distribution*. Amsterdam: John Benjamins, 2004.

Rahv, Philip. "Paleface and Redskin." *Kenyon Review* 1, no. 3 (1939): 251–56.

Rastier, François. "Computer-Assisted Interpretation of Semiotic Corpora." In *Quantitative Semiotic Analysis*, edited by Dario Compagno, 123–40. Cham, Switzerland: Springer, 2018.

Rezaie, Naghmeh. "Iranian Adaptations: Overview." *Cross Cultural Adaptations*. http://www.crossculturaladaptations.com/iranian-adaptations/overview/.

Rich Johnston. "Oblivion, Based on the Non-Existing Graphic Novel." *BleedingCool.com*, April 12, 2013. https://www.bleedingcool.com/2013/04/12/oblivion-based-on-the-non-existing-graphic-novel/.

Richards, Washna Radia. "Translating Cool: Cinematic Exchange between Hong Kong, Hollywood and Bollywood." In *Transnational Film Remakes*, edited by Iain Robert Smith and Constantine Verevis, 118–29. Edinburgh: Edinburgh University Press, 2017.

Robinson, Andrew. "An A to Z of Theory: Walter Benjamin: Art, Aura and Authenticity." *Ceasefire*, June 14, 2013. https://ceasefiremagazine.co.uk/walter-benjamin-art-aura-authenticity/.

Ryan, Marie-Laure. "Transmedia Storytelling as Narrative Practice." In Leitch, *Oxford Handbook of Adaptation Studies*, 527–41.

Said, Edward. *Culture and Imperialism*. New York: Knopf, 1993.

Saint Pepsi. "Enjoy Yourself." Illuminated Paths, 2013. YouTube audio.

Sanders, Julie. *Adaptation and Appropriation*. New York: Routledge, 2006.

Schaefer, Sandy. "Are Original Movies Really a Hard Sell?" *Screenrant.com*, June 26, 2015. http://screenrant.com/tomorrowland-box-office-original-movies-sequels/.

Schwanebeck, Wieland, and Douglas McFarland, eds. *Patricia Highsmith on Screen.* New York: Palgrave, 2018.

Schwartzman, Paul. 2016. "I Gave a Famous Rock Star a Windshield Tour of D.C.—and Didn't Know Who She Was." *Washington Post*, March 18, 2016. https://www.washingtonpost.com/lifestyle/i-gave-a-famous-rock-star-a-windshield-tour-of-dc--and-didnt-know-who-she-was/2016/03/18/8124ab28-e488-11e5-bc08-3e03a5b41910_story.html.

Scott, A. O. "Review: In Nate Parker's 'The Birth of a Nation,' Must-See and Won't-See Collide." *New York Times*, October 6, 2016. https://www.nytimes.com/2016/10/07/movies/the-birth-of-a-nation-reviewnate-parker.html.

Seigel, Robert L. "How Do You Secure the Rights for a Remake of an Old Movie?" *MovieOutline*, June 8, 2018. http://www.movieoutline.com/articles/how-do-you-secure-the-rights-for-a-remake-of-an-old-movie.html.

Seiter, Ellen, and Mary Jeanne Wilson. "Soap Opera Survival Tactics." In *Thinking Outside the Box: A Contemporary Television Genre Reader*, edited by Gary R. Edgerton and Brian G. Rose, 136–55. Lexington: University Press of Kentucky, 2005.

Self, Will. "The Novel Is Dead (This Time It's for Real)." *The Guardian*, July 18, 2013. https://www.theguardian.com/books/2014/may/02/will-self-novel-dead-literary-fiction.

Sexton, Max, and Malcom Cook. *Adapting Science Fiction to Television.* Lanham, MD: Rowman & Littlefield, 2015.

Sheffield, Matthew. "Meet Moon Man: The Alt-right's Racist Rap Sensation, Borrowed from 1980s McDonald's Ads." *Salon*, October 25, 2016. https://www.salon.com/2016/10/25/meet-moon-man-the-alt-rights-new-racist-rap-sensation-borrowed-from-1980s-mcdonalds-ads/.

Smith, Iain Robert. "*Game of Thrones*: Adaptation and Fidelity in an Age of Convergence." *Antenna*, April 9, 2015. http://blog.commarts.wisc.edu/2015/04/09/game-of-thrones-adaptation-and-fidelity-in-an-age-of-convergence/.

———. *The Hollywood Meme: Transnational Adaptations in World Cinema.* Edinburgh: Edinburgh University Press, 2017.

Smith, Iain Robert, and Constantine Verevis, eds. *Transnational Film Remakes.* Edinburgh: Edinburgh University Press, 2017.

Sobchack, Vivian. *Screening Space: The American Science Fiction Film.* New Brunswick, NJ: Rutgers University Press, 2001.

Sotiropoulou, Chrysanthi. *Elliniki kinimatografia 1965–1975 / Greek Cinema 1965–1975.* Athens: Themelio, 1989.

Spillers, Hortense J. "Mama's Baby, Papa's Maybe: An American Grammar Book." *Diacritics* 17, no. 2 (1987): 65–81.

Stam, Robert. "Beyond Fidelity: The Dialogics of Adaptation." In *Film Adaptation*, edited by James Naremore, 54–76. New Brunswick, NJ: Rutgers University Press, 2000.

———. *Literature through Film: Realism, Magic, and the Art of Adaptation*. London: Blackwell, 2005.

Stam, Robert, and Alessandra Raengo, eds. *A Companion to Literature and Film*. Malden, MA: Blackwell, 2004.

Stanhope, Kate. "*Exorcist* Remake Ordered to Pilot at Fox." *Hollywood Reporter*, January 22, 2016. https://www.hollywoodreporter.com/live-feed/exorcist -remake-ordered-pilot-at-858486.

Stengel, Richard. "Build a Monument to Capitalism." *Time*, January 11, 2002. http://content.time.com/time/nation/article/0,8599,192964,00.html.

Strong, Jeremy, ed. *James Bond Uncovered*. New York: Palgrave, 2018.

Tanner, Grafton. *Babbling Corpse: Vaporwave and the Commodification of Ghosts*. Hants, UK: Zero Books, 2016.

Tasker, Reuben. "Talking with Toyomu, the Producer Who Imagined 'The Life of Pablo' for a Tidal-Less Japan." *Genius*, April 8, 2016. https://genius.com/ a/talking-with-toyomu-the-producer-who-imagined-the-life-of-pablo-for-a -tidal-less-japan.

Thompson, Anne. "Dustin Lance Black Talks Controversial J. Edgar Script: 'We Didn't Put Hoover in a Ball Gown.'" *IndieWire*, November 23, 2011. http:// www.indiewire.com/2011/11/dustin-lance-black-talks-controversial-j-edgar -script-we-didnt-put-hoover-in-a-ball-gown-184000/.

Thompson, Ethan, and Jason Mittell, eds. *How to Watch Television*. New York: New York University Press, 2013.

Thompson, John B. *The Media and Modernity: A Social Theory of the Media*. Stanford, CA: Stanford University Press, 1995.

Thompson, Kristin. *Storytelling in the New Hollywood: Understanding Classical Narrative Technique*. Cambridge, MA: Harvard University Press, 1999.

Thornley, Davinia, ed. *True Event Adaptation*. New York: Palgrave, 2018.

Tillet, Salamishah. "'I Got No Comfort in This Life': The Increasing Importance of Patsey in 12 Years a Slave." *American Literary History* 26, no. 2 (2014): 354–61.

Tolkien, J. R. R. *The Lord of the Rings*. Boston: Houghton Mifflin, [1965] 1981.

Torgal, Luís Reis. *O cinema sob o olhar de Salazar*. Lisbon: Temas e Debates, 2001.

Toyomu. *Imagining the Life of Pablo*. Self-released, 2016. YouTube audio.

Travers, Peter. "12 Years a Slave." *Rolling Stone*, October 17, 2013. https://www .rollingstone.com/movies/movie-reviews/12-years-a-slave-113891/.

Truffaut, François. *Hitchcock/Truffaut*. New York: Simon and Schuster, 1966.

Tunzelmann, Alex von. "Is Selma Historically Accurate?" *The Guardian*, February 12, 2015. https://www.theguardian.com/film/2015/feb/12/reel-history-selma-film-historically-accurate-martin-luther-king-lyndon-johnson.

Turner, Nat, and Thomas R. Gray. *The Confessions of Nat Turner: The Leader of the Late Insurrection in Southampton, Va.* Chapel Hill: University of North Carolina Press, 2011.

VanDerWerff, Todd. "Bates Motel Has Finally Caught Up with *Psycho* and It's Glorious." *Vox*, March 5, 2017. https://www.vox.com/culture/2017/3/5/14796702/bates-motel-convergence-of-the-twain-recap-season-5-review.

Variety Staff. "Westworld." *Variety*, December 31, 1972. https://variety.com/1972/film/reviews/westworld-1200422881/.

Verevis, Constantine. *Film Remakes*. Edinburgh: Edinburgh University Press, 2005.

———. "Remakes, Sequels and Prequels." In Leitch, *Oxford Handbook of Adaptation Studies*, 267–84.

Vermeule, Blakey. *Why Do We Care about Literary Characters?* Baltimore: Johns Hopkins University Press, 2010.

Vieira, Patrícia. *Cinema no Estado Novo: A encenação do regime*. Lisbon: Colibri, 2011.

Viera, Joao Luiz, and Robert Stam. "Parody and Marginality: The Case of Brazilian Cinema." *Framework* 28, Special Issue (1985): 20–49.

Vogler, Christopher. *The Writer's Journey: Mythic Structure for Writers*. 3rd ed. Studio City, CA: Michael Wiese Productions, 2007.

Wagner, Geoffrey. *The Novel and the Cinema*. Rutherford, NJ: Farleigh Dickinson University Press, 1975.

"Washington DC to PJ Harvey: 'She's to Music What Piers Morgan Is to Cable News.'" *The Guardian*, March 17, 2016. https://www.theguardian.com/music/2016/mar/17/pj-harvey-washington-community-of-hope-six-demolition-project-ward-7.

Watts, Gerard, "Walter Benjamin's Definition of the Aura." *Digital Arts and Humanities (DAH) Structured PhD* (Royal Irish Academy), 2012. http://dahphd.ie/gerardwatts/author/gerardwatts/. Accessed April 2, 2018.

Weir, Andy. *The Martian*. London: Random House, 2014.

Wells-Lassagne, Shannon. *Television and Serial Adaptation*. New York: Routledge, 2017.

West, Kanye. *The Life of Pablo*. GOOD/Def Jam, 2016. Streaming audio.

Wheatley, Helen. *Gothic Television*. Manchester: Manchester University Press, 2006.

Whelehan, Imelda. "Adaptations: The Contemporary Dilemmas." In *Adaptations: From Text to Screen, Screen to Text*, edited by Imelda Whelehan and Deborah Cartmell, 3–18. London: Routledge, 1999.

Wickman, Forrest. "How Accurate Is Lincoln?" *Slate*, November 9, 2012. http://www.slate.com/blogs/browbeat/2012/11/09/lincoln_historical_accuracy _sorting_fact_from_fiction_in_the_steven_spielberg.html.

Wicks, Sammie Ann, "Music, Meaning, and the Adaptation of Literature." *Literature in Performance* 2, no. 1 (1981): 89–97.

Wloszczyna, Susan. "12 Years a Slave." *Rogerebert.com*, October 18, 2013. https://www.rogerebert.com/reviews/12-years-a-slave-2013.

Young, Alex. "Mount Eerie—*Pre-Human Ideas.*" *Consequence of Sound*, November 14, 2013. https://consequenceofsound.net/2013/11/album-review-mount-eerie-pre-human-ideas/.

# Contributors

**STACEY ABBOTT** is Reader in Film and Television Studies at the University of Roehampton. She is the author of *Celluloid Vampires* (2007), *Angel: TV Milestone* (2009), and *Undead Apocalypse: Vampires and Zombies in the 21st Century* (2016), and is the coauthor, with Lorna Jowett, of *TV Horror: The Dark Side of the Small Screen* (2012). She has written extensively about the Gothic and horror on film and television. Her most recent publications include articles on the history of Bram Stoker's *Dracula* on film and TV, Bryan Fuller's *Hannibal* as TV horror adaptation, and remaking the zombie apocalypse.

**KARIN BOKLUND-LAGOPOULOU** is Emerita Professor in the School of English at the Aristotle University of Thessaloniki, Greece, where she taught medieval literature and literary theory from 1981 until her retirement in 2015. Her research interests include medieval literature and semiotics, particularly narrative theory and textual analysis; she has also published work on popular culture, an interest that stems from the frequent presence of medieval motifs in fantasy literature, science fiction, film, and television. Her publications include the monograph *I Have a Yong Suster: Popular Song and the Middle English Lyric* (2002) and articles on literary theory, popular culture, and medieval literature published in Greek, European, and American journals and in collective volumes. She frequently collaborates with her husband; they have coauthored *Meaning and Geography: The Social Conception of the Region in Northern Greece* (1992) and together with Mark Gottdiener coedited the anthology *Semiotics* (2003). Their latest publication is *Theory of Semiotics: The Tradition of Ferdinand de Saussure* (in Greek, 2016).

**THOMAS BRITT** is Associate Professor in the Film and Video Studies Program at George Mason University. He is the head of the screenwriting concentration and teaches several classes, including Ethics of Film and Video. Recent publications include "'Between Two Mysteries': Intermediacy in *Twin Peaks:*

*The Return*" from *Critical Essays on Twin Peaks: The Return* and "'Came Back Haunted': Global Horror Film Conventions in *The Haunting of Hill House*" from *The Streaming of Hill House*.

**SIMON BROWN** is Associate Professor of Film and Television at Kingston University. He has published extensively on numerous aspects of film and TV, including early cinema, color cinematography, 3D, cult television, and horror. His current research centers around Stephen King and adaptation. His recent publications include an article on seriality and ABC's *Under the Dome* for *Science Fiction Film and Television* and a monograph, *Screening Stephen King: Adaptation and the Horror Genre in Film and Television*, for the University of Texas Press.

**EURYDICE DA SILVA** holds a PhD in Romance Languages, Literature, and History with a major in Portuguese Studies from Paris Nanterre University, France. Her dissertation focused on Portuguese cinema and censorship during Salazar's New State (1933–74). She holds a BA in Film and an MA in English Literature, specializing in film adaptations from the Sorbonne University, and a Certificate in Screenwriting and Producing from the UCLA School of Theater, Film and Television. With experience in the film industry at different stages of production in France, Portugal, and the United States, she is currently a screenwriter for film and television in Paris.

**JOAKIM HERMANSSON** is a doctoral candidate at Gothenburg University and teaches adaptation, narratology, and film history in the Film and TV Production Program and the Screenwriting Program at Dalarna University, Sweden. After an early career as an art curator and twenty years as a teacher, first in business and mathematics and later in social sciences and languages, he earned an MA in Linguistics and an MA in Literature. He has also been involved in research on workplace culture and has published an edited volume on art history. Today, his research interest and dissertation revolve around the representation of adulthood and social adaptation in novels, screenplays, and films.

**BETTY KAKLAMANIDOU** is a Fulbright scholar and Associate Professor of Film and Television History and Theory at the Aristotle University of Thessaloniki, Greece. She is the author of *"Easy A": The End of the High-School Teen Comedy?* (2018), *The "Disguised" Political Film in Contemporary Hollywood* (2016), *Genre,*

*Gender and the Effects of Neoliberalism* (2013), and two books in Greek on adaptation theory and the history of the Hollywood rom-com. Kaklamanidou is also the coeditor of *Contemporary European Cinema: Crisis Narratives and Narratives in Crisis* (2018), *Politics and Politicians in Contemporary U.S. Television* (2016), *The Millennials on Film and Television* (2014), *HBO's "Girls"* (2014), and *The 21st Century Superhero* (2011). She recently completed a coedited collection on post-2008 European cinema and is currently editing a volume on film adaptations. Her articles have appeared in *Television & New Media*, *Literature/Film Quarterly*, *Celebrity Studies*, and the *Journal of Popular Romance Studies*.

**URSUAL-ELENI KASSAVETI**'s research in the light of Cultural Studies has focused on different aspects of cultural theory, visual culture, visual ethnography, and popular culture (video, television, and cinema) from production to consumption in different monographs and articles. She studied Greek literature at the Athens School of Philosophy and holds an MA in Cultural Studies and an MA in History and Folklore (National and Kapodistrian University of Athens). Her PhD dissertation revolved around the 1980s VHS culture in Greece. She is an adjunct tutor in the Department of Journalism and Mass Media at the Aristotle University of Thessaloniki and in the School of Social Sciences of the Hellenic Open University.

**THOMAS LEITCH** is Professor of English at the University of Delaware. His most recent books are *The Oxford Handbook of Adaptation Studies* (2017) and *The History of American* Literature *on Film* (2019).

**NICOLE PIZARRO** is a doctoral candidate in Film and Popular Culture at The Ohio State University. She is Associate Director of the 2020 Digital Media and Composition Institute. Her scholarship focuses on adaptation as a rhetorical practice and how we adapt cultural trauma. Her work on film and comics has been featured on LatinxSpaces and The Middle Spaces. She recently received the Eric Walborn Award for Excellence in Digital Media and English Instruction.

**CHRISTINA WILKINS**, PhD, is a researcher in television and literature. She has written articles about religion, identity, and mental health in popular culture and is the author of *God Is (Un)dead* (2018), which looks at vampire figures in the post-9/11 world. She is currently working on an article on nostalgia and *Westworld* alongside a research project on mental health and contemporary literature.

Elliott, Kamilla, 10, 81, 167–68
Ellis, John, 194
Elverum, Phil, 195, 200–202
Emerson, Ralph Waldo, 28
*Exorcism of Emily Rose, The*,
     109
*Exorcist, The*, 98, 99, 101, 105,
     108–12
*Exorcist: The Beginning*, 109
*Exorcist II: The Heretic*, 109
*Exorcist III*, 109
Eyerman, Ron, 165, 167

*Fall, The*, 84
*Family Plot*, 25
fandom, 80, 82, 91
fantasy, 98, 117–18, 122, 133
*Fargo*, 97–98
Farhadi, Asghar, 21–25
*Fear the Walking Dead*, 103
Fellini, Federico, 28, 31, 47
Ferro, António, 64, 65, 67–68
Ferro, Marc, 62
fidelity, ix, 73, 81, 90, 100,
     101, 118, 120, 139, 157,
     168, 194–95, 209
Fiery Furnaces, 195, 206
*Fifty Shades of Grey*, 45–46,
     176–79, 183
figurativization, 118
Fitzgerald, F. Scott, 33
*Following, The*, 84
*Four Weddings and a Funeral*,
     54, 92
*Frankenstein*, 11, 46, 97
*From Dusk Till Dawn*, 98
*Frost/Nixon*, 49
*Futureworld*, 87–88

*Game of Thrones, A*, 103
Genette, Gérard, x, 9
*Godfather, The*, 33, 34
*Gone Girl*, 176, 178, 182, 184
*GoodFellas*, 49
graphic novel, 3–4, 42, 52, 55,
     102–4
*Great Gatsby, The*, 33
Greimas, Algirdas J., x, 9, 13,
     118, 129
Griggs, Yvonne, 2, 80
*Grindhouse*, 44
Grossman, Julie, 2, 13–14
Guimarães, Manuel, 72–73

*Handmaid's Tale, The*, 97, 98
*Hannibal*, 13, 79, 81, 83,
     85–86, 88–92
*Harry Potter*, 45, 54, 118, 132
*Harvest of Wind*, 72
Harvey, PJ, 195, 203–5
*Haunting of Hill House, The*,
     112

Hawthorne, Nathaniel, 28
*He's Just Not That Into You*, 49
*Heaven Can Wait*, 25
*Hills Have Eyes, The*, 51
*Histoires extraordinaires
     (Spirits of the Dead)*, 27
Hitchcock, Alfred, 24–25, 83,
     105–8
*Hitman*, 54
Hogan, Patrick Colm, 173–74,
     185
Holmes, Sherlock, 85
*Hope Six Demolition Project,
     The*, 203–4
*House of Cards*, 97–98
*How to Train Your Dragon*, 45
Hugo, Victor, 23
*Hunchback of Notre Dame,
     The*, 23
*Hunger Games*, 45
Hutcheon, Linda, 14, 23, 25,
     157–58, 196
hypertext, 47, 101
hypotext, 47, 101

*I Love You / S' agapo*, 145
*I Strofi / The Turn*, 145
iconography, 46, 53, 106, 140,
     150
identity, 31, 64, 66–67, 88,
     90–91, 102–5, 111, 167,
     174
*Imagining the Life of Pablo*,
     206
*Incidents in the Life of a Slave
     Girl*, 156, 168
intertextuality, 2, 48, 80, 81,
     83, 196
investigative adaptation,
     13, 93
*Iron Man*, 52
isotopy, 13, 119–22, 127–28,
     130–33
*Itazura Lolita: Ushirokara
     virgin*, 30

James, Henry, 28, 31
*Jaws*, 29
*J. Edgar*, 49
Jeffers, Jennifer L., 23, 24
Jenkins, Henry, 91, 117
Jess-Cooke, Carolyn, 2, 52
*Jimmie Higgins*, 23
Jowett, Lorna, 97

*Kaante*, 24
*Kamikaze, agape mou*, 146
*Karmen Gei*, 28
*Katiforos*, 145
*Keeping Up with the
     Kardashians*, 50
Kidel, Sam, 195, 202, 203

King, Stephen, 8, 97, 100, 112
*King's Ransom*, 24
Korda, Alexander, 23
Krallice, 193, 208

*La cage aux folles*, 23
*Lady Bird*, 5
Lang, Fritz, 24
*Last Exorcism, The*, 109
*Last House on the Left, The*,
     109
Lecter, Hannibal, 80, 83,
     86, 90
*Legion*, 109
*Lego Movie, The*, 55
Leitch, Thomas, 1, 2, 3, 4, 9,
     10, 12, 40, 47, 48, 49, 50,
     81, 82, 83, 84, 85, 86, 91,
     92, 106, 107, 185, 196
Lev, Peter, 47
Lewis, Sinclair, 28
*Life of Pablo, The*, 205–6
*Lincoln*, 42, 43
*Lisbon Song*, 68
*Lolita*, 30, 36
London, Jack, 32
Loock, Kathleen, 2, 106–7
*Lord of the Rings*, 118, 120,
     122, 127
*Losers, The*, 52
*Lost Boys, The*, 92
*Love of Perdition*, 68, 69,
     70, 71
*Love Story*, 139, 145
Lubitsch, Ernst, 25

*Machete*, 44
*Mad Max: Fury Road*, 54
*Madame Bovary*, 28
Malle, Louis, 28
manga, 44, 52
*Maré, Nossa História de Amor*,
     28
*Martian, The*, 176, 178–82
*Masoom*, 30
*May Revolution, The*, 65
*Maya Memsaab*, 28
McFarlane, Brian, ix, 1, 92
McQueen, Steve, 156, 159,
     161–62
*Me Before You*, 176–77,
     179–82, 184
mediascape, 14, 139–42, 149,
     151
*Miami Vice*, 53
*Miracle Worker, The*, 33
mise-en-scène, 46, 119
*Mission: Impossible*, 53
*Mississippi Mermaid*, 24
*Mist, The*, 98, 99
Mittell, Jason, 82
*Moby Dick*, 36

CPSIA information can be obtained
at www.ICGtesting.com
Printed in the USA
JSHW031520041120
9301JS00001B/43